The Internet for Scientists

The Internet for Scientists

Kevin O'Donnell

Scottish Agricultural Sciences Agency
Edinburgh
Scotland

and

Larry Winger

The Medical School
Newcastle-upon-Tyne
UK

harwood academic publishers
Australia • Canada • China • France • Germany • India
Japan • Luxembourg • Malaysia • The Netherlands • Russia
Singapore • Switzerland • Thailand • United Kingdom

Amsteldijk 166
1st Floor
1079 LH Amsterdam
The Netherlands

British Library Cataloguing in Publication Data

A catalogue record for this book is available from the British Library.

ISBN 90-5702-222-2 (soft cover)

We thank the contributors to Usenet News whose public posts we've snapped for
illustrative purposes; in friendly Internet spirit we sought permission for each
snapshot, and we hope we've used the ones for which appropriate permission was
received. In addition we are grateful for permission from Gustavo Glusman of
BioMOO to quote a descriptive extract; for permission from Iddo Friedberg to
quote instructions on accessing BioMOO; for permission from Barry Hardy to use
examples from the 2nd Electronic Glycosciences Conference; for permission from
Karen Traicoff to use an excerpt from 'Mars Online', part of NASA's Learning
Technology project in Association with Passport to Knowledge; for permission
from Jeanne McWhorter of Diversity University for permission to use MOO
instructions and a snapshot of DU's virtual Student Union Lounge.

Cover illustrations by Lindley Gooden B.Sc., M.Phil., P.Dip.

Contents

Meet the Authors

Kevin O'Donnell was born in Bathgate, Scotland. He graduated from the University of Aberdeen in 1984 with a microbiology degree and went on to complete a PhD at University College of North Wales in Bangor on hydrocarbon-degrading plasmids in Pseudomonas. Writing up his PhD thesis provided his first real encounter with a computer — an Amstrad PCW word-processor. Having completed his thesis, and convinced that he had exhausted all of the possibilities offered by computers, he sold his PCW. A few years later he discovered Microsoft windows and found that computers were actually fun to use and could do all sorts of useful things which made life easier. Then he gained access to the Internet and found that it was full of resources of interest to scientists but that it took a long time to find them all. Kevin O'Donnell is head of the Diagnostics and Molecular Biology section at the Scottish Agricultural Science Agency, Edinburgh. He is a member of the Society for General Microbiology, International Society for Plant Molecular Biology and British Society for Plant Pathology. He is the author of 'Molecular Diagnostics Resources on the Internet' in *Molecular Biology-based Methods in Clinical Diagnosis*, U. Reischl ed., Humana (in press).

Born near Niagara Falls, in Canada's southern Ontario, Larry Winger had a traditional rural upbringing in what could be called The Great Mid-West. Primary education in tiny, single-room red-brick schoolhouses turned into expansive secondary education in prosperous suburban Pennsylvania. Success in science projects turned into summer employment in a research lab in California, developing the earnest young scientist successively into laboratory technician, graduate student, and post-doctoral researcher. Larry learned BASIC and FORTRAN by remote teletype in 1970, spent the summer of 1971 programming statistical tests by punch card on an unappreciative mainframe, and purchased his first Apple computer (the 2 plus) in 1981, when he was a poor, struggling immunologist living in a garret in London. Although he supplemented his meager living allowance by annoying computer programmers with incredibly bad saxophone busking off Tottenham Court Road, it was not until he seduced Carrie in the dim glow of the B&W monitor that he began to realise the full potential of the personal computer. Following the short-term project trail back to Alberta's frozen wastes, the itinerant Wingers produced a daughter under the Northern lights, introduced the lab to the first IBM PC, and then abandoned it to return to England's home counties to care for ailing parents.

Here their shared projects focussed on malaria, and though the biological data was exciting, the Wingers were without any computer to share it with, having been forced to sell the 2 plus to a family of lost souls in Edmonton. A short spell in 1987 as a scientist with a biotechnology company in Geneva introduced Larry to electronic mail, as well as the horrors of weekly commuting across Europe to share in the birth of a son. Return to London meant commuting into a lab filled with Apple devotees, but confirmed a deep suspicion that computer processing is only as good as the data that goes in. It was only when the Morris Minor broke down on the way to the library, that Larry realised how convenient it was to search for citations by computer. Desperate to pay the mortgage between contracts, abstracting pharmacology papers with a bust accountancy firm's ancient computer filled a financial void in the Wingers' life. Sadly, however, it was not until three research projects later, now 1994 in clinical chemistry, that with a major purchase of a new PowerMac and full ethernet and modem connections, the Winger family became virtual. Remote and poverty-stricken now in their electronic cottage in the bleak North Pennines, they are very happy.

Foreword

A great leap in the scientific understanding of our world sometimes comes about as a result of a simple practical discovery of how something can be done, most often by one or two people. The newly unlocked door is then pushed wide open by an army of followers.

In experimental biology, my own field of interest, a good example is the methodology of cloning antibody-producing cells (Milstein and Kohler). Such discoveries are usually driven by the *frisson* that all experimentalists know, of being the first to witness how things actually work.

The army on the other hand, in which I count myself a footsoldier, may sing the discoverer's praises, but they nevertheless have an unrelated agenda, *viz.* the personal ambition to show the world how *they* have now been able to exploit the new technique.

A different kind of leap, that applies to all kinds of human endeavour, not least science, arises as a result of the discovery of any new way to communicate. Communication began with shouting and waving, but the former was improved markedly by extending its range (Marconi) and the latter by displaying the waving on a screen (Baird).

Everybody these days feels that the formal lecture is dead or dying, just like the formal printed paper with its Victorian third person singular and rigid construction. Poster sessions have enjoyed an increasing popularity. The style is informal and you only have to listen to the hum to know there is a honey-flow somewhere. The Internet strikes me as a sort of intergalactic poster session facilitating the quintessential feature of all modern communication — that of an interactive capability.

Today's IT language is undoubtedly English. "American" I hear you cry, but US-speak is much younger, so the original Anglo-Saxons can surely claim priority. Nevertheless, the USA is now the foremost cradle of Anglo-Saxon idioms and currently these derive frequently from IT (personally I am glad that scientific phoneticization withered with fosforus). So the new styles and expressions are dominated by the linguistic fashions, and these authors sit easily on that platform.

I would also describe the style of this book as "jokey", and it grew on me as I read on, although as an oldie scientist, the glossary could have been much bigger. The informality of the Internet is one of its most attractive features, and that is reflected here with all the buzzwords and humour to leaven the text commendably. Now, therefore we have a resource crucial to the army's next advance. It tells you how to get into this latest communication system in a series of blindingly simple steps.

Well over a decade ago I wrote about the depressing phenomenon of Information Overload. Until someone invents a direct connection to the human brain so that information can be fed in overnight from databanks, our poor old eyes will have to act as the restriction mechanism that keeps us sane. You cannot read it all, you know. The idea of this book is that it tries to address the problem of how to cope with an army who have discovered a whole new toy shop full of goodies, and they are all shouting and waving about it. If bedlam and babel are not to ensue and drive us all mad, we must have some order. A communication system that will reach the widest possible audience has the best chance of stopping the nightmare of rediscovering what is already known.

It is likely that the authors will wish, and will be pressed, to update and expand this work on an almost continuous basis but I feel they are up to it. There will be CD versions and similar presentations for other specialist groups. Eventually of course everything will end up on the Internet itself. For scientists, a well-managed world neural network should lessen the chance of re-inventing the wheel every three years, and we should be grateful that someone is addressing orderliness so soon.

Professor Arnold R. Sanderson
– Department of Immunology, Royal Free Hospital
– General Secretary British Society for Immunology, 1979–1986

Acknowledgements

KO'D would like to thank his wife Elaine, who put up with a backlog of his household chores while this book was being written, and his son Euan, who now knows the answer to the question "Mummy, who's that strange man who lives in the room with the computer?"

LW recognises with thanks the support received from macinfo@ncl, and stimulating discussions on the daily commute with his good mate Andy Morgan, whose succinct advise, "Finish the job, and stop whining," was instrumental in the completion of this book.

KO'D and LW both wish to thank Lindley Gooden for his inspired cover illustrations. They also acknowledge Harwood for support and encouragement at every stage of the project.

INTRODUCTION

An overview of the purpose of this book, and what can scientists can do with the Internet

WHY ANOTHER BOOK ON THE INTERNET?

There are now dozens of self-help books on the Internet, as everybody, their uncle and various other extended family members are joining the excited rush to get on-line. Few, if any, of these books, useful as they are to the recreational, net-surfing public, are particularly good guides for working scientists. That's an odd oversight, since the net was first set up by scientists for scientists. This book aims to correct that oversight, by providing a straightforward manual and review handbook for accessing and using the resources of the Internet in the day to day labours of the working scientist. Of course, if you are surfing the Internet already, and know your way around telnet, FTP, the WWW, newsgroups and mailing lists, then our recommendation is either (a) move along to the really useful chapters in which scientific resources in each of these areas are considered, or (b) buy this book for a colleague who has not yet seen the light.

WHO THIS BOOK IS FOR

This book was written for people like us. We are two scientists who have learned through trial and error how to find and use the resources available for scientists on the Internet. We don't know very much about computers, we see them very much like our cars; our interest lies in using them to get from A to B, not in fiddling around underneath the bonnet or avidly reading technical specifications. There is a phenomenal amount of information on the internet — and at least 90% of it is rubbish. In writing this book, we have sifted through the rubbish to allow you to go directly to the resources which will help you in your work. The Internet is now just so vast, with its own idiosyncrasies and lacking any central index, that the hard-working scientist might not have enough time to explore its potential.

WHAT THE INTERNET CAN DO FOR YOU

Many readers of this book will have seen press reports about the Internet and will have heard colleagues talking about it. Most likely you will have access to a PC (note that we include Apple Macs, Acorns, etc. in our generic term PC) and you will have an Email address supplied by your university or company. This book will help you exploit these tools to the full, while taking only a minimal amount of time away from the important work at the bench, or in the classroom, or the serious task of grant and paper writing. Let us consider a few examples of what we mean when we say that the Internet can be a useful tool in the successful scientist's armoury.

- A scientist goes into work. The new method has failed to work again — what could be going wrong? A mail message is sent to an audience of thousands of other scientists asking if anyone else has had the same problem. Within minutes, replies start arriving. The advice you get saves you wasting days.
- You have money in your budget to buy a new piece of equipment, but you don't know which company to buy it from. You put a message on an Internet newsgroup asking for recommendations and receive several posts from colleagues commenting on their purchasing decisions.
- You have sequenced a gene and want to compare it with other sequences. You go to a web site and launch a similarity search using software on a computer on another continent, and receive an answer within minutes.
- You want to do a bit of background reading on a subject but it isn't covered by your department's small library. You initiate a search for appropriate WWW sites using a programme on a distant computer. You receive a selection of WWW addresses for resources devoted to this subject — including an on-line tutorial.
- You may have colleagues who can already use the Internet to do all of the things we've mentioned so far in the time it takes you to open your day's in-tray. Perhaps you would like to be able to do these things yourself — but where do you start?

Start right here. By the time you have read this book you will either know where the resources appropriate to you are or know how to look for them. You will have joined the tens of thousands of colleagues who are already making use of the Internet as a matter of routine. Like any new resource, you may be unsure of how to use it and get the most out of it. This book will enable you to make your way through the pitfalls and help you to use, and contribute to, science on the Internet.

No matter how obscure your field, you will be able to communicate with colleagues doing similar work all over the world. All of a sudden, the world is a small place. It is as easy, and as quick, to swap documents and data with colleagues across the globe as with one across the corridor. The development of contacts and spread of information previously carried out by the slow process of face to face communication at conferences and correspondence generated by a publication has now been speeded up by some orders of magnitude. This is transforming the face of science as it makes collaboration so much easier and more fruitful. At the same time, those scientists who do not take part in this information revolution are in danger of being left behind. They will be as cut off from their colleagues as they would be if they didn't have a telephone or a postal address.

DO ANY OF THESE SCENARIOS SOUND FAMILIAR?

Let's look at some practical examples in more detail, using some composite scientific characters:

Scenario 1:
Jane Q. Sirius has just landed her first lecture-ship at a red-brick university. She produced excellent, topical work on her thesis in Paleobiology, and went two steps further on her postdoc in the field with a famous big name. She has to gather her wits about her now, get her collection of papers and references suitably collated and cross-filed, write as many grant applications as she can, send off those two papers on the last several months' field-work that have been put off during the move, prepare the lecture notes for the second year's stimulation, serve as the department's seminar co-ordinator, and the faculty's library committee — oh, and be a mother to her two boys. She's tried accessing the Internet from her ethernetted computer (whatever that means), but the information from the university computing service is in another language, and frankly, she doesn't have the time to get into it. Besides, she's managed just fine, to date, thank you very much, and the only real kudos will be for the money she brings in, by virtue of the real papers she publishes.

Scenario 2:
John St. John Realitas is a powerful professor and head of his department of Environmental Anthropology. The other members of the group are all deferentially in awe of him, and so they should be, considering his track record. He's got a contact list as long as a gorilla's arm and he works it, hard, at several conferences a year. The department is fortunate

to be solidly placed in the field, thanks to his team's work (well, his work, if the truth were known), and he's responsible for its steady source of income. He has heard about some of the anthropological discussion groups available on the Internet, but they're really for the striving, overeager junior members of the department, he supposes. But what was that bright young postdoc talking about the other day, a web-presence? Maybe he should look into that. But getting the smart-alecky computer service boyoes up again for a show-and-tell is really more than he can bear. He's already got dozens of unanswered email messages sitting quietly in his computer, and frankly he wishes the information revolution would just slow down and let him catch up.

Scenario 3:
Nease S. Androj is working on his first postdoc after finishing the thesis. It was a hard go, finishing off that damned thing, and there are still several papers to be got out for the advisor, but just now it's a new lab, a 'slightly' different subject (plasma, as opposed to quantum physics), and 4 other postdocs who are more eager, more brilliant, and apparently more together than Nease. They're always talking about what they've retrieved from the net, and of course the latest electronic papers are *de rigeur* for discussion at the weekly lab meetings. Nease gets them okay, now, thanks to the friendly help of one of the students, and reads them on print-outs just before the 4pm gatherings. But wouldn't it be nice to have something unique and interesting for the meetings? Something one could call one's own? 'Hmmmm, I wonder if anybody's working on my sort of question? How can I find out without broadcasting my own, or heaven forbid, the lab's research secrets all over the world? More to the point, if I set up my computer at home, can I still get around the net as easily as from the institute? Why does everybody always assume that physicists are naturally computer-literate?' Nease has got away without letting the secret slip out, by assuming diligent desk-work, with arcane formulae, and seriously heavy thoughts, but it's going to be obvious to all before long that something is being hidden. Yes, Nease is ever-so-slightly computer-phobic, and how will that play with the references for the next posting?

Scenario 4:
Alistair M. Filbert has taught science in the local secondary school for the past 15 years. He's got the lesson plans in his head, and frankly, it hasn't been too much trouble coping with the new curriculum. When you've done it all before, well, you've got a pretty good reservoir of experience to call on, in emergencies. Over the years, some of Mr.

Filbert's students have done pretty well in the Science Festivals — once one of them got into the National Festival! Of course, you don't get a chance to do any real research at secondary school level, though wasn't that 16 year old lad's Nature paper on the chaos of a dripping tap marvellous! Now *he* had a dedicated teacher behind him! But you've got to keep thinking ahead, even (especially) if you want to get (keep) your head of department rank, and Mr. Filbert is not immune to the massive media hype of the Internet. The Internet could be the way to go. But the Technology department doesn't seem too keen, though the English teachers are wetting themselves, practically, with glee over the latest net gossip. It's a mad, weird world, and to be honest, if he could just see what some of the possibilities for school science are, without losing too much time from the evening's marking, then he could decide how to proceed. But how to get started? Alistair's nephew tried to show him some cool sites, but he wasn't too impressed with the Oasis home page, nor the spelling on the alt.fan.xyz newsgroup. How can he find out whether or not it's worth the effort?

You wouldn't be reading this book, of course, if we didn't think that there really is a great potential for enhancing useful science work, by appropriate use of the communication tools of the Internet. That is to say, we wouldn't have written it!

Each of the internet tools will be described in its own chapter. This description will be followed by a list of resources which you can access by using the knowledge gained from the chapter, putting it to work for you straight away before you have a chance to forget it. These are resources which have been designed by scientists for scientists and have been accessed by us in the course of writing this book.

Remember, we're both working scientists, as it says on the back, and although we've been involved with the Internet for the past few years, we still have our daily bread and butter work to accomplish. And we realise that we're only just beginning to learn how to use the Internet creatively, and seriously, in our daily working lives, so we do apologise for any errors that may have crept into the information herein. Such knowledge as we have, however, we'd like to share. That's part of the Internet spirit, by the way (BTW — Internet speak is full of TLAs).

If you discover that some information has changed, needs adding to or even (very unlikely of course) that something is just wrong, then we hope that you will enter into the Internet spirit by submitting your correction to our WWW page. This will not be an on-line version of the book — otherwise you wouldn't buy it — but will be a sort of continually updated on-line erratum slip.

LET'S CONSIDER THOSE EXAMPLES AGAIN:

1. **For the busy Jane Q. Sirius, into her first real lectureship, what is the point of the Internet?**
    Well, Jane will want to expand her contact base, to include new friends and acquaintances from around the world. Of course, primary research, publishing and grantsmanship are still her primary interests and responsibilities, but she can keep her ear to the ground, and continue to watch the field develop by listening in on public and semi-public discussion groups in her field. She could use some advice on how to minimise waiting, and maximise her time investment in the Internet.

2. **What can the Internet bring to Prof. Realitas, that he hasn't already got through his contacts?**
    If the professor's department still doesn't have a smart web site, then it is just not with it, frankly, and it is beginning to look rather over-the-hill and tired. Where are the professor's next generation of students coming from? The brightest of next year's crop will have spent a good part of their secondary school time canvassing their potential university on the Internet. Those departments that can show the most exciting ideas, with useful hypertext linkages, will be the departments that will attract the most eager of students. And how can Prof. Realitas dramatically demonstrate the work of his entire department, unless he brings a constant stream of visitors through for a tour past the accumulated posters hung outside the labs? The Prof. needs a WWW site, and he needs to get his faculty members jumping, collating their poster information into computer images, and exciting text, so that the work of his dynamic department is always on display to the world's commercial and academic interests.

3. **How can Nease Androj improve communication skills, impress the Prof., and come up to par with the senior postdoc peers?**
    Anybody can be a lurker. You don't have to broadcast your presence when you read the publicly displayed queries and responses in the various newsgroups. Nor do you have to respond to all of the messages in the discussion groups whose members are connected by a list server that conveniently contacts each and every one, with each and every message. Nease can sit, think, consider and reflect, and slowly, gradually, at a very personal pace, begin to communicate some questions. First in private communications back and forth to trusted friends, then in semi-public discussions, and finally in public newsgroups.

When Nease has found the needed self-confidence, through interactions with other Net-workers, it will be less difficult to broach questions and attempt responses in the very real setting of the lab meeting. Thank goodness the unpublished results of that colleague in Japan diverted him from that blind alley just in time!

4. **Can Mr. Filbert ever catch up? And how much will it cost?**
   Sadly, it is always a question first of finances, and what the budget will allow; almost the equal of financial cost, though is the price of time.

   On the money front, there is good news for Mr. Filbert: new initiatives from governments, local education authorities, and interested computer companies may mean that hook-up can be cheaper than he had dared hope. But how much time will he have to spend to navigate his way around the ftps, the telnets, the arcane data bases and special pleaders, before he finds something really useful for his science classes? Relax, Mr. Filbert, yes there are really useful resources out there waiting to be utilised, (like lesson plans and teaching aids) for early (Key Stage 3) and late secondary (Key Stage 4), as well as some lovely sites of special interest to your colleagues at the primary level. What's more, there are forums where you can communicate with teachers just like you who are learning to use the Internet in their lessons.

We aim to provide a worker's approach to the Internet. We don't guarantee that we will cover all the important sites, and we expect our work here to date rapidly; such is the nature of the evolving Internet. What we do hope to achieve, however, is a simple and succinct approach to using the tools and resources of the Internet, so that science workers can quickly and appropriately enhance their own work, and understand the basic operative procedures. Surely (abject apologies here and for the future for calling you Shirley, if you are a Charlotte or a Hank) that's a useful ambition, and we would be much obliged if you'd let us know, after reading the book and putting its suggestions to the test, whether or not we've achieved that goal.

But enough of our goals and objectives; how did this thing called the Internet come about, anyway?

ALL THE HISTORY OF THE INTERNET THAT YOU *REALLY* NEED TO KNOW

The precursor of the Internet was ARPAnet, an experimental network of computers designed to withstand nuclear attack. Scientists at the different sites covered by the ARPAnet soon discovered the usefulness of electronic communication and began to write the software that would enable the net to function. The first non-military extension of this was the establishment of 5 super-computer centres at sites across the USA. Other US institutions were connected to these super-computer centres by regional networks, forming a chain by which any institution could be connected to any other. This was funded by the NSF, who made one of the conditions of funding that access had to be made widely available, leading to an explosion of people with Internet access. At around the same time the technology was being developed to connect computers in a local area network or LAN, a collection of ethernet-connected computers sharing resources, for example printers, and able to communicate with each other. The internet is a series of networks connected together, so that any computer on any part of the constituent networks can communicate with any other computer and share resources. However, instead of allowing you to share a printer with half a dozen other users, the Internet allows you to share messages, files and software with millions of other people.

Now we are not going to waste too much of our time and yours discussing the physical parameters of initial hook-up. Except to note that if you are at home, you will need a modem, first of all, and if at the lab, then no doubt you are already ethernetted (direct linked to your local network of computers), and are being taken care of by your obliging computing service.

There is no way around the hardware acquisition, unfortunately, and this is a time and financial question that each of our exemplary characters (and you) will have to deal with in their (your) own way. You have got to spend your or your institution's money and time getting the ethernet or the modem. Look at it this way — compared to the investment in the personal computer, a modem or an ethernet connection has got to be peanuts, both in time and money. Now we ourselves know that ethernet connections are hard-wired into the computer, and so data transfer via ethernet is only dependent on the capacity of your Department's system, and we also understand that modems work down the phone line, but that is about the extent of our computer knowledge. In most ways, you see, we are only working scientists, like you, and we are barely computer-literate. But our eyes are open and we would like to open yours too.

We believe, however, that you should be able (like we have done, for home use) to buy a good modem from any reasonable computer store (and when we say good, we realise that 14.4k V42bis is the minimum you will want to use nowadays, and if you can get a good deal on a 28.8k, we are envious — but see our "getting connected" comments, below.)

Alternatively, there are lots of different modems advertised in your friendly computer magazines, and they'll be happy to take your order over the telephone. If you are connecting from work, it should only take your computing service people half a day to wire in an ethernet terminal connection, assuming your building is connected. But we will leave these practicalities to you and your service. We want to deal with what you do next. Okay, we'll try to contain our enthusiasm and take the step-wise route.

Right, let's assume that the physical hardware is in place, and you have got the software installed as per your computing service, or thanks to your friendly local service provider.

The service provider is your (and our) link to the Internet. Service providers could be your university or institute, or they could be a commercial outfit specialising specifically in providing service, via servers, naturally enough. We don't really want to mention any, since there are quite a few now, and uncanny mention of one might offend the others. But as we write, the computer magazines (especially the ones dedicated to the Internet, like *Internet* or *.net*) are full of ads from specialist service providers. Between us, we use 4 different service providers (a research institute, 2 different universities, and a commercial provider).

If you are not affiliated with an institute, university, or industrial research concern with its own Internet server, then it is a simple matter of enlisting the services of your local service provider, by paying some money up front as well as a monthly rate, usually, and getting a server to act as the link between you, your computer, the modem, and the good old Internet. The thing to remember, as you consider the financial commitment here, is that you've got to pay your telephone company for the telephone time you use, as well as pay the server fees to the service provider.

HOW DOES IT WORK? AS MUCH TECHNICAL INFORMATION AS YOU WILL EVER NEED

The key to modern internet communications is the TCP/IP protocol. It is tempting, but misleading, to think of a computer connection as being like a phone connection — a constant, if temporary, link between two computers. However, if this was the case the network would soon freeze up due to the resources wasted keeping open a connection down which no information was being exchanged. In fact, information exchange takes place on a quantum rather than wave model i.e. information is exchanged in discrete packets rather than in one continuous stream.

Information is sent from one location to another according to IP (internet protocol). This involves putting the data to be transferred into packets of up to 1500 characters in size. A header is attached to the packet giving the numerical address of the destination computer. As it is quite likely that much of the data that needs to be transferred will be larger than 1500 characters, a given message or file needs to be broken down into a series of packets. This is done according to the transmission control protocol (TCP). This numbers each packet so that they can be pieced together again in the right order at the other end. It also adds a checksum, so that the receiving computer can check that all the data was received — if the checksum doesn't add up it requests that the data is resent.

All this happens every time you use the internet — but it happens so quickly and smoothly you will be unaware of it. The process will be carried out by your TCP/IP software 'under the bonnet'. One piece of information you will need to add if you are sending data is the numerical address of the destination computer. This is made up of 4 numbers joined by dots. Each number is known as an octet and the full address is known as a dotted quad. However, these numerical addresses themselves work 'under the bonnet'. What you will use are the more memorable addresses made up of words. For example, Kevin O'Donnell's numerical email address is 193.130.50.22. Fortunately, there is no need for anyone to use this number (indeed, he had to go and look it up to put it in this intro-duction). Instead we can use odonnell@sasa.gov.uk, which is a much more logical address (to we humans at least) and easier to remember. The computers know that these names correspond to the underlying numeri-cal addresses — so we never have to.

The addresses are constructed according to the domain name system. Each of the octets corresponds to a domain or level in the address, corre-sponding to the levels in a postal address, with the highest level domains corresponding to the country the computer is in. The domain name system means that any computer can find another by forwarding the message on to the next level. This is explained in more detail with regard to mail messages in the email chapter but the principle holds good for all infor-mation traded between Internet computers.

GETTING CONNECTED

Now really, in this day and age, most readers of this book will have a 'free' connection provided by your university or company, either in the form of an Internet connection to a computer in your own laboratory or office, or to a central computing facility which is used by a larger number of people.

In either case, the local network administrators (sysadmins, or sysops, as they are known in Internet jargon) will maintain the ethernet connection between your computer and the rest of the local network and possibly provide some of the software too. If, however, you are in the position of having to figure out your own connection (as for example, from home), then there are a number of companies which do this. As we have mentioned, your connection will require a personal computer, a telephone line and a modem (generally, the faster, the better and certainly not slower than 14,400 bps). Computers maintained by university sysadmins have a permanent connection to the Internet, via a dedicated high-speed line and a router, a computer which acts as an interface between the local network and the Internet.

With a commercial service provider your computer only becomes connected once you dial up to your Internet service provider. You will then be on-line and your computer will have become part of the Internet. This connection is via a computer at the service provider. The methods of connection are SLIP (serial line internet protocol) and PPP (point to point protocol). The details of these two things are unimportant. What matters is that they are a way of allowing home based PCs to access the Internet via an ordinary phone line and modem. Such facilities are commonly referred to as SLIP/PPP accounts.

When you open an account with a service provider, you will likely be given the software necessary to make TCP/IP connections — TCP/IP stack software such as Trumpet Winsock. You will also be given the software necessary to send Email, download files using the File Transfer Protocol and to browse the World Wide Web. All of this software must be installed on your own PC. The act of dialling up your SLIP/PPP service provider doesn't allow you to access software at the other end of the line — the service provided is the transfer of data to and from the Internet site that your PC has become on dialling in. Your modem connection is likely to be slower than the ethernet connection for permanently on-line machines, however you will have the advantage that it is your personal account and you are not therefore beholden to your employers for the use you make of it. It is becoming increasingly common for people with an Internet connection at work to have an account with an Internet service provider for their home computer for this reason.

One drawback to this personal freedom, is that you will have to configure your network software yourself, a task carried out for you on work-based PCs by sysadmins. But honestly, this is not in itself a very complicated task. A good service provider will ensure that full instructions are given to you on setting up your connection and it is really just a matter of following them. A good service provider will also have a help-line that you can call on for support if you run into difficulties. The speed of the connection is dependent upon two factors: the maximum speed of your modem, and the speed at which the service provider can handle your data. There is no advantage in, for example, having a 28,800 bits per second modem if the highest connection speed for your provider is 9600.

The other big difference between a home connection and a connection at work is that it is *you* that has to pay! The amount and system of billing differs widely between providers. Some charge a flat rate per month. Others charge according to the time spent on-line. The latter is likely to be more expensive but you are also less likely to be kept waiting by an engaged tone. In both cases, you will also have to pay for your phone calls to the provider. For this reason it is a good idea to find one with a local POP (point of presence). This is a local phone number that you use to connect to the service, meaning that your connection time is charged at the local rather than long-distance rate.

Follow the instructions from your service provider carefully and explicitly, note the help line (try to call at off-peak times; yes, easier said than done), and keep persevering until you get the magic connection accomplished. Sometimes connections are set up smoothly and sometimes all sorts of problems can crop up. The variety of combinations resulting from differences in hardware, software and Internet providers means that it would be pointless, not to mention impossible, for us to deal with all of them within the scope of this book. So we're not going to deal with any. The people who will help you are the help desk at your Internet provider. Remember these words of comfort: whatever problem you have, someone else less intelligent than you has had it already and it got solved. You've got to get the connection, and you've got to follow the instructions. We can't help you until you're really actually ready to fly!

So you've taken the commercial service provider route, and frankly, it has taken you an hour after dinner to install the software bits and pieces on your computer, and you've lost your temper crawling around under the table placing and replacing the modem connection, and the clip to the phone line. But now you've actually dialled in, the software has clicked, whirred, buzzed and jangled, and you suddenly get a message somewhere intimating you're connected with something! What next?

Or alternatively, the computing service technician from sysadmin has just installed all the software on your machine at work, and smiling at you, avers that it's all ready to go, maybe even shows you how to double click on one of the icons, and then says good-bye. What do you do now?

Yes, you are ready to begin your working adventure on the Internet. We're only too pleased to show you the way to these world-wide resources. This book will tell you what they are, how to reach them and how to use them. We will deal with communication forums such as electronic mailing lists and usenet newsgroups of specific interest to scientists. We will also deal with accessing on-line courses, information, software and databases via the World Wide Web. As well as showing you how to access the increasing amount of information, we will show you how to contribute to the virtuous cycle by making your own information available on the Internet.

The Internet continues to grow so rapidly that there is always a chance — in fact a certainty — that some of our information will be out of date by the time you get to read it. But you can check it out our on our own home pages, on the WWW, for up to the moment, spot-on accurate info. And if we're not correct, then you can email us immediately, and let us know. We'll put the correction on our web page for others to see — that way you'll be helping others and have made your first contribution to the virtuous circle that is Science on the Internet.

Kevin O'Donnell – odonnell@sasa.gov.uk
Larry Winger – larry.winger@newcastle.ac.uk

Internet for Scientists web site:
<<http://www.compulink.co.uk/~embra/ifs.html>>

COMMUNICATING WITH OTHER PEOPLE

Email

Dr Jonathan Fairlamb, **efficient research scientist at his govern-ment's National Institute for Standards, receives his mail every-day in his pigeon hole outside the main office. He spends a good hour every morning with his correspondence. For some of it he writes out replies in longhand which he sends to the typing pool to be turned into official looking letters of the type that were sent to him. What with holidays and someone off ill, the typing back-log is now up to 2 weeks, so Professor Realitas will have to wait for a reply to his invitation to give a seminar at his department next month — but what's the alternative? Other pieces of corre-spondence he puts into his intermediate in-tray where it soon gets buried under mounds of other non-urgent correspondence and quietly forgotten about. With email, it need not be like that. Dr. Fairlamb is really within almost immediate touch of his clientele. At his computer, receiving electronic mail from enquirers, he could highlight immediate questions of concern, answer salient points, and make his own, on a reply message that includes whatever of the original query he wishes. Furthermore, both query and reply are conveniently filed away on his personal computer, for later reference. For letters which absolutely must go out on the official letterhead, he can now send his draft to the typing pool by email, allowing them to make the necessary modifications very quickly.**

INTRODUCTION

Electronic mail, or email (pronounced E-mail) as it is always called, is the simplest electronic communication tool at the scientist's disposal. Essen-tially it is a means of sending messages from one computer terminal to another, which might be in another part of the university or in another part of the globe. Internet sceptics often make the point that they have ad-equate means at their disposal already for sending written messages — the postal service or, in case of urgency, the fax machine. However email is more convenient for a number of reasons.

Firstly, it is faster to compose and to send. For example, an email message might arrive in your in-tray from the editor of a journal, asking you if you would be willing to referee a particular paper. By clicking the reply button on your email software, you can send off a reply straight away just by typing in a brief response and clicking another button to send the message. Not only will this be quicker in terms of the time it takes for the message to reach its recipient, it will be quicker in terms of the amount of your time it will use up — and more convenient than a telephone call. Secondly, it will be cheaper — an email to anywhere in the world will cost as much as the connection time to send the message — virtually nothing. Thirdly, and most importantly, the versatility of email extends far beyond sending simple text messages to a single recipient. The capability exists to join in email conferences or mailing lists on subjects of particular interest. Modern email software has the capability to send not just text messages but also images, spreadsheets — in fact any file that exists on your computer. If you don't have access to other Internet tools, you can also use email to find and retrieve files from remote computers.

In this chapter we will describe how email can be used, the basic features of email software, email mailing lists and other things that can be done by email. The chapter concludes with a directory of current scientific mailing lists.

EMAIL: HOW IT WORKS

In essence, email works like ordinary postal mail. A message is sent from your address to another address where the addressee will open it and read the contents. Virtually every scientist and science student has the facility to send and receive email to anyone with an email address. If you have used any part of the Internet at all, it will have been this — even if it's just to swap Christmas greetings with that colleague you met at a conference last year.

The first thing to look at here is email addresses. A lot of information can be gathered from these and this helps to explain how the system works. Take one of ours, for example:

odonnell@sasa.gov.uk

Like postal addresses, email addresses start with a specific person and progress to more general parts of the address. This is because the Internet works just like the post office in the way it delivers mail. There are now millions of email addresses. If you are on a computer in Australia then your own host computer is unlikely to know who 'odonnell' is — unless there is an 'odonnell' on your local network, which is not a lot of help if it's not the one you want to contact. Likewise 'odonnell@sasa' will draw a blank. However, odonnell@sasa.gov might give more of a clue — he obviously works for the government — but which government? The full address, odonnell@sasa.gov.uk gives the computer the information it needs to send the mail. It looks at the last part 'uk' and forwards the message to a computer in the UK. The UK computer knows where to send 'gov.uk' messages, the gov computer knows where sasa is and the sasa computer knows that 'odonnell' is a user on its system and delivers the message to Kevin O'Donnell's in-tray. As you will have sent it in the middle of the night (his time), he can read it first thing the next morning.

The part of the address to the right of the '@' sign is known as the fully qualified domain name (FQDN). As well as 'uk' there are domain abbreviations for all other countries with email capability. The only exception to this is the USA, where although '.us' is possible, it is not common. This happens for the same reason that UK postage stamps are the only ones not to have the name of the country on it i.e. the USA was the first place to have email, therefore the addresses had no need to state the obvious by tagging '.us' on the end. Endings found on the end of US addresses can provide information about the nature of the place where it originates

.com = commercial
.edu = education
.gov = government
.mil = military
.net = network organisation
.org = non-profit organisations

So it *is* possible to derive some information from email addresses, even if you are only a human.

EMAIL SOFTWARE

There is not space enough in this book to describe all of the email software packages available. However, we will describe some of the features which are common to most of the packages available today — some of which are available as Freeware (for explanation of 'Freeware' see the FTP Chapter).

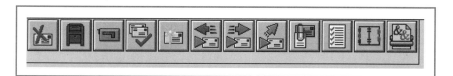

In essence, their operation is very simple. You click on a button that creates a form for you to use to send a message. There will be a 'To' field where you type in the recipient's address, a 'cc' field where you can add addresses that you would like to copy the message to, a 'Subject' field where you can add a title for your message and a field where you can type in the message itself, just as if you were using your word processing programme. However, unlike a word processing programme, you are restricted to plain text messages without things like bold, italics, different typefaces and so on.

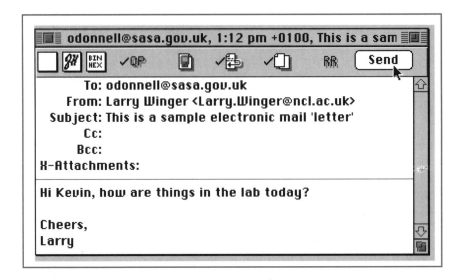

As well as the ability to create, address and send messages your email package should have some or all of the following features.

1) **An address book.** Instead of looking up and typing in addresses by hand every time you want to send mail, you can simply open up the electronic address book by clicking on the appropriate button, select the address(es) you want and these will be entered into the 'To' part of the email message.

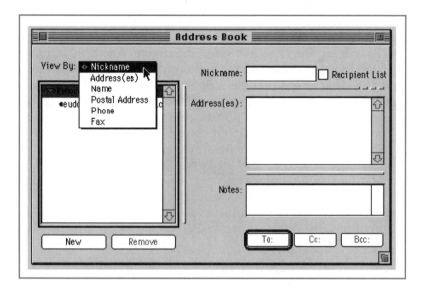

2) **Folders.** As you accumulate email messages you will want to store them in a way that makes them easy to find if you wish to refer back to one. Most packages will allow you to create different folders, just like in a word processing package, so that you can keep messages from different mailing lists (see below), departmental and social messages separate.

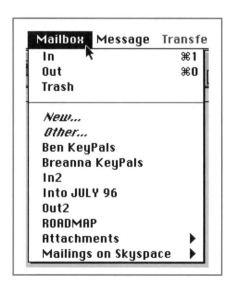

3) **Attachments.** This makes email incredibly powerful. The ability to send attachments means that you can not only send a message to a colleague saying 'I've got some interesting results' you can also attach the spreadsheet file that shows them – all at the click of a button. This means that the recipient will open your message and, if interested, click on the attachment file to open it. As it opens, it will start up the appropriate spreadsheet software (in this instance) on your colleague's computer. Of course this will only work if your colleague's computer has software compatible with your own. It isn't a lot of use to send a Word for Windows file if the recipient is using a Macintosh word processing programme that can't read it and vice-versa (in case you're ever in that position one possible thing to do is to convert the file into plain ASCII text; another is to find out specifically from your recipient what translation capacity they have).

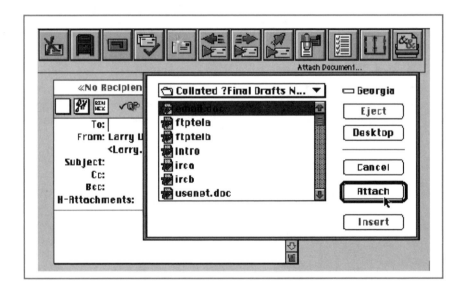

4) **Spell checker.** You are only likely to find this on commercial software packages. It can make a real difference to the quality of the messages you send out.

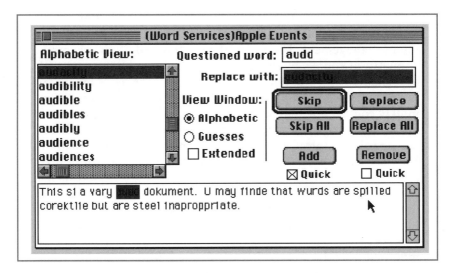

5) **Mail filtering.** Incoming mail from different locations is automatically directed to designated folders. This is useful if you are subscribed to several mailing lists and wish to keep the posts separate.

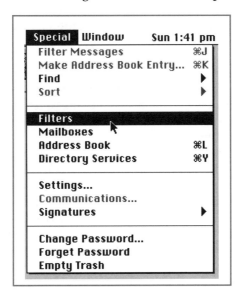

MAILING LISTS

Sylvie Amorjam, **consulting dietician for the National Health Service, must keep up to date with developments in her field, as well as develop new research areas for study and continue to prove the point of her posting, to an increasingly budget-conscious management. The national conference of dieticians meets annually, of course, but budgets are considered by the new rank of managers on a quarterly basis, and today the news media are jumping on a hyped-up episode of salmonella outbreak in remotely cooked hospital meals. How can Sylvie demonstrate that her diets are fresh and wholesome, and that she is responding to the ongoing clinical crisis with current research information? How are her colleagues handling the situation? She can call her colleagues, but each call takes some time, of course, and tears her away from her own work – not to mention the inconvenience to the dietician on the other end of the phone line. What is needed is to have that dieticians conference going on *right now*, so that a concerted, professional response to immediate crises can be mounted quickly, and so that the best advice is available when she needs it — but of course that would be impossible wouldn't it? Actually, no it wouldn't. What Sylvie Amorjam needs is an email subscription list for like-minded individuals, so that at the stroke of a single message for help, all of the 500 or so health scientists interested in nutrition in her country, as well as those across the world, are alerted to the problem and can join in discussions on dealing with it.**

The most obvious use of email is to send messages to one other individual and we have already seen how this can be more convenient than other methods. However, email also offers the potential to communicate with more than one person at a time – sometimes hundreds at a time. This is done through mailing lists.

Email mailing lists are run by several different software packages, each of which runs on the same basic principle. The best-known and most widely used mailing list software is called Listserv and this has become something of a generic term. In essence, any mail sent by a member of a mailing list to the list address will be automatically forwarded to everyone else on the list. Each mailing list has two addresses. One is the automated listserv address, which will be in the form of listserv@a.server.somewhere. This address will handle a few commands which allow subscribers to join or leave the list and sometimes obtain other data such as addresses of subscribers or archive files. Each listserv programme may be handling several lists at once so it is necessary to specify the list you wish to join or leave. For example, here is the message that Charles Darwin would send to subscribe to the creation-science list run by the listserver at fossils.pah.edu (the actual wording may vary slightly with different types of software but we will deal with those later):

To: listserv@fossils.pah.edu

Message: subscribe creation-science Charles Darwin

The listserv software would receive this message, automatically extract Charles Darwin's email address from the header of the message and add him to the list of subscribers for the creation-science list. As the listserv can only understand a limited number of commands, it is important to leave the subject line blank and not to add any signature to the message, otherwise you will receive back a message from a stressed computer consisting of "Kevin — command not recognised, O'Donnell — command not recognised . . .". Yes, we've been there.

An automated message would then be sent to Mr Darwin containing information about how to post to the list, how to unsubscribe etc. It is important to keep such messages. As you will discover, it can be irritating to find a series of messages in your in-tray sent to the hundreds of people on the list containing subscribe/unsubscribe commands. This distinction between the automated address for sending messages to the software and the address for sending messages to everyone on the list is an important one, not only because you need to know it to get it to work but out of courtesy to the other subscribers.

Having successfully subscribed to a list what will you find? It varies enormously.

Date	K	Subject
10:07 pm -0400	2	Re: Unlabelled Specimens -Rep
8:22 pm -0600	2	Re: Unlabelled Specimens
8:07 pm -0400	2	PREDICTIONS!!!
11:54 pm -0500	2	Re: Unlabelled Specimens
11:11 pm -0700	3	Re: PNI: Re:Psychological facto
11:43 pm -0700	14	New List 10-17-96
10:12 am +0100	4	Schoolscience Digest U3 #92

2474/7152K/53K

Each list has its own internal culture built up out of in-jokes, old arguments, prominent members and so on. In some ways it's like being a member of any club, so it's always a good idea to read posts for a while and get the lie of the land before getting right in with your own posts. It's generally true to say that people are more relaxed and informal in email communications than they are face to face. Considering that, in many cases, members of a list are complete strangers, it is striking how quickly 'relationships' build up between the members. It is quite a difficult thing to describe — the best thing to do is to join one and follow the discussions. You need never actually contribute anything — and indeed many never do — but, in a list devoted to your own specialist subject, eventually something will come up where you will feel you want to make a contribution. An allied question that net-novices often ask is "how useful is it?". The answer to this depends to an extent on you.

An email mailing list is like any other club or society, it is what its members make it. If there is a frank and friendly exchange of views and data then the list can become very useful. If you receive quick and accurate answers to questions related to your field then you will come to see the list as part of your work. However, this can only work if people give as well as take. If somebody posts a question that you know the answer to, then you should contribute by sending in a response.

The typical way that lists work is that someone will post a question "What do people think about technique X/reagent Y/Z's recent paper". Other subscribers will comment – " I tried that method, and the trick is to . . ." – and still more will follow up those comments and so on. It can sometimes feel as though people are having real-time conversations. The level of information is best characterised by saying that it matches the sort of conversations that take place over the coffee break at conferences, rather than during the conference sessions themselves.

As with face to face conversations there are certain rules of etiquette that should be adhered to. This is all the more important because although if you are having a series of exchanges with one person it can seem as though you are having a one to one conversation, you are in fact communicating with every member of the mailing list. Most of these people will never have met you and will judge you on how you come over in your posts. This is somewhat less of a problem with mailing lists which are relatively small, specific and cosy, than with newsgroups, (see next chapter) which have a potential audience of millions. Nevertheless, it pays to take some time over your contributions and make sure that they are grammatically correct and contain no spelling mistakes. You should never get drawn into personal attacks, no matter how abusive the other person is; subscribers to a list quickly get tired with both parties and will forget that you have right on your side. Most importantly, if a list is restricted to a specific topic, respect that. If there is a list designed for communication between creationists, they will not appreciate you pointing out why they are wrong. You may be right but you are also being rude; there are other forums for arguments of this kind if you wish to pursue them. In addition, if you post quotes of Richard Dawkins to a list for creationists, then they might retaliate by posting chunks of biblical text to the neo-darwinist list you enjoy reading so much. The net result would be everyone ending up with posts in their email in-tray that they didn't want to read, the opposite of what was intended. So, respect the rules of the game. If you really want an argument you will have no problem finding one on Usenet (see Newsgroup chapter).

If you find a list that you think sounds interesting, for example, if you are interested in marsupials and find a marsupia-l in the directory at the end of this chapter, how should you proceed? First you should find out a bit more about the list by requesting any information about it from the appropriate listserver with the "info marsupia-l" command. This will give you some information about the list, if the list owner has created any. You could also try the "review marsupia-l" command, which will give you a list of the subscribers to the list, if the owner has chosen to make this publicly available. If not the listserver will tell you the total number of subscribers.

If the information told you that marsupia-l was a list for people who thought that wallabies were cute and cuddly then you might surmise that the scientific content was likely to be low. If, on the other hand, the information file told you that marsupia-l was for scientists interested in the study of marsupial biology, you would know that it was a serious list. Likewise, the review command will have supplied some useful information. If you find that the list has only 30 subscribers it is likely to be a fairly low volume one, if it has 1,000, then it is likely to be busy, with several posts landing in your in-tray each day. You can then make a decision as to whether it is likely to be useful for you to subscribe to. It might even be possible for you to retrieve some old posts, if these have been archived. If this is the case, you will be sent instructions on how to retrieve archived posts when you subscribe.

Once you decide to subscribe to a list, you will start to receive messages. It is best to wait a few days and see what 'threads' of discussion are going on before making your own contribution. For example, if you joined a group of people already chatting at a conference, you wouldn't immediately start up a topic of conversation, you would listen to the conversation that was going on when you joined, and then contribute if appropriate. The same is true of email conversations.

Properly used, mailing lists can be a good source of expert information. However, as with all Internet communications, there are few quality checks on information except your own good sense. Lists are also under the control of one or more list-owners (also called moderators) who are typically the people who went to the trouble to set up the list in the first place. It is their job to stimulate discussion, intervene if exchanges are abusive or inappropriate posts are made. The exception to this are moderated mailing lists. These are lists in which the posts are not immediately sent to all of the subscribers but are first sent to the moderator who checks to see if they are appropriate for the list. This has the advantage of reducing junk mail but the disadvantage of increasing the time it takes for posts to appear and therefore information to be exchanged. Moderated lists are still the exception rather than the rule but the increasing volume of junk email — chain letters etc. — appearing on mailing lists may change this. One of the drawbacks of increased access to the Internet is that there is a proportional increase in the number of idiots and a disproportionate increase in the amount of damage they can do.

As mailing lists are publicly available, even those which have fairly esoteric scientific subjects will attract a few interested lay-people. Opinion on this varies. In Bionet lists, and many others, it is made clear that the list is for use by professional scientists and not for answering homework questions, for example. Our own opinion is that as we scientists always complain about the public's lack of interest in science, we should take the time to help those who are displaying an active interest. Generally speaking this is a low level problem on mailing lists — the real nuisances prefer to embarrass themselves in front of millions on the newsgroups. However, a special word of caution needs to be added for medical professionals. People suffering from a particular illness will seek out information about it on the Internet and will find relevant mailing lists (they might look at this book for example). For this reason many medical lists require subscribers to submit some information about themselves first. Even if this is the case, a good rule of thumb is the following: do not put any information about a patient in an email that you would not be happy to write on a postcard. Mail can be intercepted and read at any of the computers your mail passes through to get to its destination. The chances of anyone wanting to do this are small, of course. However this is one reason for the growing interest in encrypting mail, a subject on which we will say more later.

SUBSCRIBING TO MAILING LISTS

There are a number of ways to subscribe to mailing lists. Listed below are the main types of programme and the basic commands that they recognise for subscribing and unsubscribing to mailing lists.

ALMANAC

To subscribe to an almanac list, send the command

> **subscribe** *name of list*

To leave the list

> **unsubscribe** *name of list*

Sending help will retrieve a full set of commands for almanac servers.

LISTPROC

To subscribe to a ListProcessor list, send the command

> **subscribe** *name of list your name*

To get information about the list, send the command

> **information** *name of list*

To leave the list

> **signoff** *name of list*

A full list of listproc commands can be obtained by sending the help command.

Listproc was designed to be similar to listserv and, confusingly, some servers use the name 'listserv' for subscribing etc. This can cause problems because although the command for subscribing is identical, the commands for unsubscribing and retrieving information, amongst others, are different. Such servers are indicated in the following table however it is a good idea to keep the information about unsubscribing and other commands that you will receive when you join a list.

LISTSERV

To subscribe to a list, send the command

> **subscribe** *name of list your name*

To leave a list send the command

> **unsubscribe** *name of list*

To get a full list of current commands send the command

> **help**

MAILBASE

To subscribe to a list send the command

> **join** *list name your name*

To unsubscribe to a list send the command

> **leave** *list name*

To retrieve a list of current commands, send

> **help**

MAJORDOMO

To subscribe to a list, send

> **subscribe** *name of list*

To leave a list, send

> **unsubscribe** *name of list*

To retrieve a list of current mailing lists maintained on a particular server

> **lists**

To retrieve a list of current commands, send

> **help**

MAILSERV

> **subscribe** *name of list*

> **unsubscribe** *name of list*

> **help**

A list of scientific mailing lists, arranged by subject, can be found at the end of this chapter.

SETTING UP YOUR OWN MAILING LIST

It may well be that you have looked around for a marsupial list and been unable to find one. You know from your work that there are a lot of scientists around who would like to talk to each other about marsupials and you know how useful mailing lists can be for that purpose. So you decide to set one up, but how do you go about it?

There are several ways of setting up a mailing list. The simplest is if your workplace already has listserv software installed or is willing to install it at your request. It is then a simple matter of asking them to set up the marsupial list and for you to publicise its existence amongst those who might be interested.

If this is not possible then there are organisations which will set up mailing lists on your behalf.

BIOSCI

BIOSCI will set up a mailing list for you on any biological science topic for a trial period of 6 months. After that time, all scientists who use the bionet newsgroups can vote on whether the list should be made permanent. Although it will still be available as a mailing list if it passes, it will also be a Usenet newsgroup. For that reason the bionet hierarchy is dealt with in more detail in the next chapter. For our purposes here it is enough to say that if you have a proposal to set up a biological mailing list you can send your proposal to biosci-help@net.bio.net. You will be required to draw up a charter stating the purpose of the list, to publicise it and to stimulate discussion on it. However administration of the list will be handled for you by BIOSCI, free of charge. In addition all posts are stored in a searchable archive, making this a valuable resource for biologists.

Mailbase

The Newcastle-based National Mailbase Service will set up mailing lists for anyone with an 'ac.uk' address (apart from undergraduate students). The reason for the 'ac.uk' restriction is that Mailbase is funded by the Joint Informations Committee of the various UK Higher Education Funding Councils. For the same reason, lists which have a high proportion of non-UK subscribers are technically frowned upon, though in practice this restriction appears to be widely ignored (and indeed it is difficult to see how it could be enforced).

Aside from the 'ac.uk' qualification, all that Mailbase ask is that your proposed list should benefit the UK higher education and research community. This means benefit their work, not benefit them by creating a list to discuss *The X-Files*. In return Mailbase will set up and administer the list, archive the posts and also any attendant files that list members might want to download. You are responsible for promoting the list and generating discussion.

Posts should conform to the JANET Acceptable Use Policy which, in common with other areas of the Internet, declares that unsolicited advertising and other commercial posts are unacceptable.

To get more details on starting up a Mailbase list visit the WWW site at http://www.mailbase.ac.uk (see World Wide Web Chapter) or send an email to mailbase-helpline@ mailbase.ac.uk

DIY

Another alternative is to run mailing list software on your own machine. Modern email programmes, and dedicated mailing list software, make this a possibility. The problem is that they will only work when your machine is connected to the Internet. If used with a home-based PC, messages will only be received and re-distributed when you dial up to your service provider. Some internet providers offer a way around this by running Listservs, which subscribers can rent space on for their mailing lists. That way your list will operate happily even when you are off-line.

The last, least sensible method of running a mailing list is to do it by hand. This means that you have to manually add and remove people from the subscription list and to manually re-distribute the messages to them. This is only really suitable for low-volume lists where time delays between message receipt and transmission aren't a big problem. In our experience such lists can be frustrating and seldom fail to give the impression of being a bit of a botched job. If this is not the impression you want to give, do not attempt to run one in this way. What *can* work well in this format is an electronic newsletter. This has no pretensions of being an interactive form of communication but is simply a regular publication of articles, distributed in electronic rather than printed form.

OTHER THINGS TO DO WITH EMAIL

You can do a surprising amount with email — some of which, however, is less convenient than the Internet tools created specifically to do the jobs, and which we describe in another chapter. However, if you don't have these Internet tools available, you can use email to locate and obtain software packages available over the net. You can even use email to obtain WWW pages. We don't have the space to tell you about these here but luckily there is somewhere you can get the information from for free. You can retrieve a copy of Bob Rankins *Accessing the Internet by Email* by sending an email message (what else?) as follows:

If you live in the US, Canada or South America, email the address

mail-server@rtfm.mit.edu

with the message:

send usenet/news.answers/internet-services/access-via-email

If you live anywhere else in the world, email the address

mailbase@mailbase.ac.uk

with the message

send lis-iis e-access-inet.txt

This document is very helpful if all you have is email access. For the rest of us however, life is much easier, as the rest of this book will show.

Index to Scientific Mailing Lists

So, you've got the software, you've read this chapter and you'd like to get stuck in to a few mailing lists — but you don't know where to start for your subject. We suggest that you start here, with this reasonably up to date list of scientific mailing lists. There are on-line indexes of mailing lists but none specifically for science, so save time with this — and remember to let us know of any that we've missed.

How to Use this Index

The mailing lists are divided into broad subject categories and then listed in alphabetical order of list name. To subscribe to any list, send a message including the list name to the listserv address given for it. The exact syntax will depend on which of the mailing list programmes mentioned previously is being used to run the list. For example, if the listserv address is majordomo@an.example.edu, then use the commands listed under majordomo.

Discipline	List Name	Subject	Listserv address
Agriculture	ag-exp-l	Agricultural expert systems	listserv@vm1.nodak.edu
	ag-lit	Issues related to agricultural literacy	listserv@tamvm1.tamu.edu
	agnews	Agriculture news Releases	listserv@vm.cc.purdue.edu
	agric-econ	Agricultural economics	mailbase@mailbase.ac.uk
	agric-l	Agriculture discussion	listserv@uga.cc.uga.edu
	agroecol	Agroecology	listserv@wvnvm.wvnet.edu
	anmgt-l	Animal management — research and practical applications	listserv@unlvm.unl.edu
	beef-l	beef specialists	listproc@listproc.wsu.edu
	biophysical	Sustainable agricultural and rural development	majordomo@tempo.undp.org

Discipline	List Name	Subject	Listserv address
	camase-l	Quantitative methods of research on agricultural systems and the environment	listserv@nic.surfnet.nl
	carlu	Comtemporary agricultural and rural land use	listserv@ukcc.uky.edu
	cofarm	Agricultural research	listserv@umd.umd.edu
	cwaenet	Women in agricultural economics	listserv@ers.bitnet
	d-mgt	Dairy management	listproc@listproc.wsu.edu
	dairy-l	Dairy discussion for professional educators and extension workers	listserv@umdd.umd.edu
	dairynew	Newsletters pertaining to dairy production	listserv@umdd.umd.edu
	dlo-e-info	Information on Dutch research on agriculture, nature and environment	listserv@nic.surfnet.nl
	dom_bird	Domesticated/Farmyard bird discusions	listserv@plearn.edu.pl
	dssat	Crop models and applications	listserv@uga.cc.uga.edu
	ecol-agric	Ecological agriculture	mailbase@mailbase.ac.uk
	ext-meat	Meat specialists	listserv@vm1.spcs.umn.edu
	exttqm	TQM milking and dry cow	listserv@psuvm.psu.edu
	extvet	Extension veterinarians	listserv@unlvm.unl.edu
	farm-mgt	Farm management	listserv@vm1.nodak.edu
	food-law	Laws dealing with food science	listserv@vm1.spcs.umn.edu
	forecon	Forest economics	listproc@listproc.wsu.edu

Discipline	List Name	Subject	Listserv address
	goats	Goat management	listproc@listproc.wsu.edu
	gp-bcfor	Sustainable development in B.C. forestery	listserv@nic.surfnet.nl
	gsa	Genetic stock administrators' discussion	listserv@iubvm.ucs.indiana.edu
	h-rural	Rural and agricultural history	listserv@msu.edu
	harma-pi	Agricultural informaticians from the EU periphery	listserv@irlearn.ucd.ie
	harma-ps	Agricultural statisticians from the EU periphery	listserv@irlearn.ucd,ie
	hih-l	Human issues in horticulture	listserv@vtvm1.cc.vt.edu
	ipfanr-l	International Programme for Food, Agriculture and Natural Resources	listserv@nervm.nerdc.ufl.edu
	ipm-pmw	Integrated pest management issues of the Pacific Northwest	listproc@listproc.wsu.edu
	manrrs	Minorities in agriculture, natural resource and related science	listserv@tamvm1.tamu.edu
	miffs	Michigan integrated food and farming systems	listserv@msu.edu
	mulch-l	Exchange of information on mulch-based agricultural systems	listproc@cornell.edu
	newcastle-disease	Newcastle disease in poultry	mailbase@mailbase.ac.uk
	newcrops	Discussion of new crops	listserv@vm.cc.purdue.edu
	nubeef-l	Beef production	listserv@unlvm.unl.edu
	pig-med	Pig production medicine	listserv@vm1.spcs.umn.edu
	rusag-l	Russian agriculture	listserv@umdd.umd.edu

Discipline	List Name	Subject	Listserv address
	sanrem-l	Sustainable agri-culture and natural resource management	listserv@uga.cc.uga.edu
	sheep-l	Sheep	listserv@listserv.uu.se
	sustag-l	Sustainable agriculture	listproc@listproc.wsu.edu
	swine-l	Journal of Swine Health and Production	listserv@vm1.spcs.umn.edu
Anthropology	anthro-l	anthropology	listserv@ubvm.cc.buffalo.edu
	asaonet	Oceanic Anthropology	listserv@uicvm.uic.edu
	h-sae	Society for Anthro-pology of Europe	listserv@msu.edu
	jwa	Journal of World Anthropology	listserv@ubvm.cc.buffalo.edu
	natchat	Aboriginal peoples discussion	listserv@tamvm1.tamu.edu
	native-l	Aboriginal peoples	listserv@tamvm1.tamu.edu
	pan-l	Physical anthropology news	listserv@freya.cc.pdx.edu
	radanth-l	Radical anthropologists	listserv@american.edu
Archaeology	amino-acid-dating	Amino acid dating of fossils	mailbase@mailbase.ac.uk
	arch-l	Archaeology	listserv@tamvm1.tamu.edu
	arch-metals	Archaeo-metallurgy	mailbase@mailbase.ac.uk
	arch-theory	Archaeological theory in Europe	mailbase@mailbase.ac.uk
	britarch	Archaeology in the UK	mailbase@mailbase.ac.uk
	c14-l	Radiocarbon dating	listserv@listserv.arizona.edu

Discipline	List Name	Subject	Listserv address
	dddnet	Taphonomy and other fossil preservation issues	listserv@uicvm.uic.edu
	dinosaur	Dinosaurs and other archaeosaurs	listproc@lepomis.psych.upenn.edu
	gisarch	Archaeology and GIS	mailbase@mailbase.ac.uk
	histarch	Historical archaeology	listserv@asuvm.inre.asu.edu
	micropal	Micropaleontology	listproc@ucmp1.berkeley.edu
	pacarc-l	Pacific Rim archaeology	listproc@listproc.wsu.edu
	palclime	Paleoclimate, paloecology for the late Mesozoic and early Cenozoic periods	listserv@sivm.si.edu
	paleobot	Paleobotany	listserv@listserv.dartmouth.edu
	paleolim	Paleolimnology	listserv@nervm.nerdc.ufl.edu
	paleonet	Paleontology network	listserver@nhm.ac.uk
	sub-arch	Underwater archaeology	listserv@asuvm.inre.asu.edu
Astronomy	apchem	Astrophysical chemistry newsletter	mailbase@mailbase.ac.uk
	funastro-l	Interesting new discoveries in astronomy	listserv@nervm.nerdc.ufl.edu
	hastro-l	History of Astronomy	listserv@wvnvm.wvnet.edu
	space	Digest of sci.space.tech newsgroup	listserv@uga.cc.uga.edu
	space	Space	listserv@umslvma.umsl.edu
	space	Space digest	listserv@ubvm.cc.buffalo.edu
	space-sh	Digest of sci.space.shuttle newsgroup	listserv@uga.cc.uga.edu
	spacenws	Digest of sci.space.news newsgroup	listserv@uga.cc.uga.edu
	spacepol	Digest of sci.space.policy newsgroup	listserv@uga.cc.uga.edu

Discipline	List Name	Subject	Listserv address
	spacesci	digest of sci.space. science newsgroup	listserv@uga.cc.uga.edu
Biology	amphibian decline	Biology and conservation of declining amphibians	listproc@ucdavis.edu
	aquarium	fish and aquaria	listserv@emuvm1.cc.emory.edu
	asz-l	American Society of Zoologists	listserv@cmsa.berkeley.edu
	batline	Bat research	listserv@unvma.unm.edu
	bcregs-l	Biological control regulations	listserv@uafsysb.uark.edu
	bee-l	Research and information concerning bee biology	listserv@uacsc2.albany.edu
	bio-hlth	Biomedical research/ healthcare	listserve@asuvm.inre.asu.edu
	bioacoustics-l	Bioacoustics	listproc@cornell.edu
	biodiv-l	Biodiversity networks	listserv@ftpt.br
	bioguide	A Biologists Guide to Internet Resources	bioguide@yalevm.ycc.yale.edu
	biojobs-l	Jobs for biologists	majordomo@mtu.edu
	biomat-l	Application of all types of materials in biology and medicine	listserv@nic.surfnet.nl
	biomch-l	Biomechanics and human or animal movement discussion	listserv@nic.surfnet.nl
	biomoo-announce	Biomoo announcements	listproc@ripken.oit.unc.edu
	biomoo-talk	Biomoo discussions	listproc@ripken.oit.unc.edu
	biophy-l	Biophysical discussions	listserv@ ubvm.cc.buffalo.edu
	bioregional		listproc@csf.colorado.edu

Discipline	List Name	Subject	Listserv address
	biorep-l	Biotechnology Research in the European Union	listserv@hearn.nic.surfnet.nl
	biosafty	Biosafety discussions	listserv@mitvma.mit.edu
	birdband	Bird bander's forum	listserv@listserv.arizona.edu
	bugnet	Insects	listproc@listproc.wsu.edu
	camel-l	Camel research	listserv@sakfu00.bitnet
	cbio		majordomo@brahms.amd.com
	cell-cycle	Cell cycle biology	mailbase@mailbase.ac.uk
	class-l	Classification and phylogeny	listserv@ccvm.sunysb.edu
	class-l	Classification, clustering and phylogeny estimation	listserv@ccvm.sunysb.edu
	cnidaria	Cnidarian biology research	listserv@uci.edu
	community-bio		majordomo@tempo.undp.org
	comphear	Comparative and evolutionary biology of hearing	listserv@umdd.umd.edu
	confocal	Confocal microscopy	listserv@ubvm.cc.buffalo.edu
	cstb	Canadian Society for Theoretical Biology	majordomo@biome.bio.dfo.ca
	cybsys-l	Cybernetics and systems	listserv@bingvmb.cc.binghamton.edu
	cze-itp	Problems of capillary	listserv@vm.ics.muni.cz
	darwin	Darwin and natural selection in today's society	listserv@yorku.ca
	elasmo-l	American Elasmobranch society	mailserv@umassd.edu

Discipline	List Name	Subject	Listserv address
	entomo-l	Entomology	listproc@uoguelph.ca
	env-klamath	Klamath bioregion discussion	majordomo@igc.apc.org
	equine-immunology	Equine immunology	mailbase@mailbase.ac.uk
	et-w4	Biological nitrogen fixation	listserv@searn.sunet.se
	ethology	Animal behaviour and behavioural ecology	listserv@searn.sunet.se
	ex-sci	Exercise science research forum	listproc@ucdavis.edu
	eye-movement	Eye movement studies	mailbase@mailbase.ac.uk
	faces-l	Interdisciplinary study of faces	listserv@utepvm.ep.utexas.edu
	farm-bio		majordomo.tempo.undp.org
	fmdss-l	Forest management DSS	listserv@pnfi.forestry.ca
	glyco-l	Glycobiology	listserv@list.nih.gov
	herp-l	Herpetology	listproc@xtal220.harvard.edu
	humbio-l	Human biology interest group discussion list	mailserv@acc.fau.edu
	immuno-logy-vacancies	Immunology job vacancies	mailbase@mailbase.ac.uk
	insect-physiol	Insect physiology	listproc@msstate.edu
	itrdbfor	Dendrochronology forum	listserv@listserv.arizona.edu
	leps-l	Lepidoptera	listserv@yalevm.cis.yale.edu
	lpn-l	Laboratory primate newsletter	listserv@brownvm.brown.edu

Discipline	List Name	Subject	Listserv address
	mammal-l	Mammalian biology	listserv@sivm.si.edu
	mollusca	Mollusc evolution and taxonomy	listproc@ucmp1.berkeley.edu
	morphmet	Biological morphometrics	listserv@cunyvm.cuny.edu
	neuropsych	Neuropsycholog	neuropsychology-request@ mailbase.ac.uk
	nose	Nasal research	mailbase@mailbase.ac.uk
	ots-l	Organisation for tropical studies	listserv@yalevm.cis.yale.edu
	oxnetbiol	Oxygen deficiency response	mailbase@mailbase.ac.uk
	parasite-genome	Computing in parasite genome research	mailbase@mailbase.ac.uk
	phospho-lipids	Phospholipids	mailbase@mailbase.ac.uk
	physiology	Physiology	mailbase@mailbase.ac.uk
	primatology	Primatology	mailbase@mailbase.ac.uk
	pupil	Pupil anatomy and physiology	mailbase@mailbase.ac.uk
	rmbl-l	Rocky mountain biological laboratory	listserv@umdd.umd.edu
	simuliidae	Simuliid (blackfly) biology	mailbase@mailbase.ac.uk
	socinsct	Social insect biology research	listserv@cnsibm.albany.edu
	taxacom	All matters relating to biological systematics	listserv@cmsa.berkeley.edu
	uv-b-research	Effect of UV-B radiation on organisms	mailbase@mailbase.ac.uk
	visres-cortical	Involvement of the cortex and cortical centres in vision	mailbase@mailbase.ac.uk

Discipline	List Name	Subject	Listserv address
	worgnet	Women Ornithologists Research Group	listserv@sivm.si.edu
	zoo	Artificial zoological society	listserv@sjsuvm1.sjsu.edu
	zoognus	News from the National Zoological Park, Washington DC	listserv@sivm.si.edu
Biotechnology	biotech	Biotechnology	listproc@gmu.edu
	biotech	Biotechnology	listserv@umdd.umd.edu
	biz-biotech	Commercial applications of biotechnology	listserv@netcom.com
	brg-forum	Bioencapsulation Research Group Forum	listserv@ciril.fr
	env-biotech		majordomo@igc.apc.org
Chemistry	catalyst-centre	Heterogeneous and homogeneous catalysis and chemical engineering	mailbase@mailbase.ac.uk
	chemchat	Chemistry discussions	listserv@uafsysb.uark.edu
	chemconf	Chemisrty research and education	listserv@umdd.umd.edu
	chemcord	Chemistry laboratory and programme coordination	listserv@umdd.umd.edu
	chemdisc	Chemconf discussions	listserv@umdd.umd.edu
	cheme-l	Chemical engineering	Listserv@ulkyvm.louisville.edu
	chemic-l	Chemistry in Israel	listserv@vm.tau.ac.il
	chemtech	Chemical technology	listserv@miamiu.muohio.edu
	chminf-l	Chemical information sources	listserv@iubvm.ucs.indiana.edu
	cho-data	Carbohydrate spectral data	listserv@uga.cc.uga.edu

Discipline	List Name	Subject	Listserv address
	cicourse	Chemical information courses on the internet	listserv@iubvm.ucs.indiana.edu
	cmts-l	Chemical management and tracking systems	listserv@cornell.edu
	corros-l	Corrosion special interest	listserv@listserv.rl.ac.uk
	electro-chemistry	Electrochemistry	mailbase@mailbase.ac.uk
	gt-atmdc	Atmospheric dispersion of chemicals	listserv@nic.surfnet.nl
	hiris-l	High resolution infrared spectroscopy	listserv@icineca.cineca.it
	hpprot	HP protein chemistry users	listserv@tamvm1.tamu.edu
	indicate	American Chemical Society Indicator	listserv@sjuvm.stjohns.edu
	inorganic-chemistry	Inorganic chemistry	mailbase@mailbase.ac.uk
	matls-l	Materials synthesis	listserv@psuvm.psu.edu
	nrcce-l	National Research Cente for Coal and Energy	listserv@wvnvm.wvnet.edu
	oligosac	Oligosaccharins	listserv@uga.cc.uga.edu
	pad	Polymer analysis and characterisation	listserv@listserv.syr.edu
	polyphenols	Science of polyphenolic compounds	listproc@ucdavis.edu
	sg-applied-catalysis	Catalysis in chemical reactors	mailbase@mailbase.ac.uk
	xafs-spectroscopy	X-ray absorption fine structure spectroscopy	mailbase@mailbase.ac.uk
Education	aaae	American Association for Agricultural Education	listserv@vm.cc.purdue.edu

Discipline	List Name	Subject	Listserv address
	aets-l	Association for the Education of Teachers in Science	listserv@uwf.cc.uwf.edu
	appl-l	Computer applications in science and education	listserv@vm.cc.uni.torun.pl
	atidvr-l	Research co-ordination in science education software	listserv@technion.technion.ac.il
	biocis-l	Biology curriculum innovation study	listserv@sivm.si.edu
	biology-teaching	Teaching of biology in higher education	mailbase@mailbase.ac.uk
	biopi-l	Secondary Biology Teacher Enhancement PI	listserv@ksuvm.ksu.edu
	c-edres	Educational resources on the Internet	listserv@unbvm1.csd.unb.ca
	cbadm-l	Computers in university biology education	listproc@liverpool.ac.uk
	chem-education	Chemistry education	mailbase@mailbase.ac.uk
	chemcom	Chemistry education in the community	listserv@ubvm.cc.buffalo.edu
	chemed-l	Chemistry education	listserv@uwf.cc.uwf.edu
	conslt-l	Consultation and discussion of research and practice in mentoring	listserv@iubvm.ucs.indiana.edu
	distlabs	Teaching science labs via distance	listserv@indycms.iupui.edu
	edstat-l	Statistics education discussion	listserv@jse.stat.ncsu.edu
	edudeaf	Deaf education	listserv@ukcc.uky.edu
	engineering-geotech-mtgs	Teaching of geotechnical engineering	mailbase@mailbase.ac.uk

Discipline	List Name	Subject	Listserv address
	entree-l	Environmantal training in engineering education	listserv@nic.surfnet.nl
	fish-junior	Knowledge transfer between marine scientists and children/high school students.	listserv@searn.sunet.se
	gentalk	genetics forum for high school teachers and students	listserv@usa.net
	geo-courseware	courseware for Earth Sciences teaching	mailbase@mailbase.ac.uk
	geocal teaching	Debate on geography	mailbase@mailbase.ac.uk
	hec-l	Higher education consortium for mathematics and science	listserv@uwf.cc.uwf.edu
	hesca	Health sciences education	listserv@listserv.dartmouth.edu
	iapadv	International arctic project adventure	listserv@iubvm.ucs.indiana.edu
	iasee-l	Solar energy education	listserv@vm.gmd.de
	iet-inst	Interactive educational technologies:instruction	listserv@uhupvm1.uh.edu
	iet-tech	Interactive educational technologies: technology	listserv@uhupvm1.uh.edu
	ifpe	Psychoanalysis and education	listserv@kentvm.kent.edu
	imse-l	Institute for math and science education	listserv@uicvm.cc.uic.edu
	itforum	Instructional technology	listserv@uga.cc.uga.edu
	jcmst-l	Journal of Computers in Mathematics and Science Teaching	listserv@vm.cc.purdue.edu

Discipline	List Name	Subject	Listserv address
	jei-l	Technology in education	listserv@umdd.umd.edu
	jse-announce	Journal of statistics in education announcements	listproc@jse.stat.ncsu.edu
	jse-talk	Journal of statistics in education discussion	listproc@jse.stat.ncsu.edu
	jte-l	Journal of Technology Education electronic journal	listserv@vtvm1.cc.vt.edu
	jvre-all	Journal of virtual Reality in Education complete journal	listserv@maelstrom.stjohns.edu
	jvre-toc	Journal of Virtual reality in Education table of contents	listserv@maelstrom.stjohns.edu
	maths-itt	Maths initial teacher training	mailbase@mailbase.ac.uk
	maths-support	Non-traditional maths teaching	mailbase@mailbase.ac.uk
	meconet	Medical education consortium network	listserv@cms.cc.wayne.edu
	medical-education	Medical education	mailbase@mailbase.ac.uk
	medical-it	IT in medical education	mailbase@mailbase.ac.uk
	mindson-net	Discussion about science teaching	majordomo@igc.apc.org
	models-sciteched	Modelling in science and technology education	mailbase@mailbase.ac.uk
	mserg-l	Maths/science education research	listproc@listproc.gsu.edu
	msupbnd	Math science upward bound discussion list	listserv@ubvm.cc.buffalo.edu
	narst-L	National Association for Research in Science Teaching	listserv@uwf.cc.uwf.edu

Discipline	List Name	Subject	Listserv address
	ncprse-l	Reform discussion list for science education	listserv@ecuvm.cis.ecu.edu
	nsela-l	National Science education Leadership Association List	listserv@uwf.cc.uwf.edu
	ntwforum	New technologies in physics education	listserv@indycms.umn.edu
	phys-l	Forum for physics teachers	listserv@uwf.cc.uwf.edu
	physhare	Sharing resources for high school physics	listserv@psuvm.psu.edu
	physlrnr	Physics learning research	listserv@idbsu.idbsu.edu
	psa-ed	Psychoanalysis and education	listserv@kentvm.kent.edu
	science-education	Study of science education	mailbase@mailbase.ac.uk
	scimat-l	Arkansas science and math education	listserv@uafsysb.uark.edu
	sebsel	NCBES Science, Engineering, Business and Science Education	listserv@listserv.arizona.edu
	smart	Science/math teacher training list	listserv@uriacc.uri.edu
	susig	Teaching in the Mathematical Sciences With Spreadsheets	listserv@miamiu.acs.muohio.edu
	tlths	Teaching and learning technologies for the health sciences	listserv@wvnvm.wvnet.edu
	twt	Teaching with technology	listserv@yorku.ca
	virtual-dissection	Computer-based teaching of anatomy and histology	mailbase@mailbase.ac.uk

Discipline	List Name	Subject	Listserv address
	water-ed	Education professionals conducting outdoor aquatic science	litserv@ukcc.uky.edu
	yescamp	Youth engineering and science camps of Canada	listserv@unbvm1.csd.unb.ca
	iams	Screen-based resources in university science education	mailbase@mailbase.ac.uk
Engineering	anemo-l	Anemometry in all kinds of turbulent fluid flows	listserv@nic.surfnet.nl
	bhs-hydrology	Hydrological science and engineering	bhs-hydrology-request@ mailbase.ac.uk
	caeds-l	Computer-aided engineering design	listserv@listserv.syr.edu
	civil-l	Civil engineering research and education	listserv@unbvm1.csd.unb.ca
	eman	Manufacture of electronic products	mailbase@mailbase.ac.uk
	engineering-concrete	Concrete research	mailbase@mailbase.ac.uk
	engineering-geotech	Geotechnical engineering	mailbase@mailbase.ac.uk
	engsoc-l	Engineering societies	listserv@listserv.unb.ca
	eppd-l	Engineering and public policy discussion list	listserv@unbvm1.csd.unb.ca
	euearn-l	Eastern Europe telecommunications	listserv@ubvm.cc.buffalo.edu
	ev	Electric vehicles	listserv@sjsuvm1.sjsu.edu
	geodesic	Buckminster Fuller's works	listserv@ubvm.cc.buffalo.edu
	hvac-fdd	Heating, ventilation and air conditioning systems science	mailbase@mailbase.ac.uk

Discipline	List Name	Subject	Listserv address
	maes-l	Society of Mexican American Engineers and Scientists	listserv@tamvm1.tamu.edu
	mech-l	Mechanical engineering	listserv@utarlvm1.uta.edu
	merrp-l	Minority engineering recruitment and retention programme	listserv@uicvm.uic.edu
	nsbeline	National Society of Black Engineers	listserv@listserv.syr.edu
	sampe-l	Society for the Advancement of Materials and Process Engineering	listserv@psuvm.psu.edu
	sci-eng	Science enginers list	listserv@gu.uwa.edu.au
	steam-list	Steam generators, piping and equipment	listproc@mcfeeley.cc.utexas.edu
	tss-list	Transportation science	listserv@mitvma.mit.edu
	two-stroke-forum	two-stroke combustion engines	mailbase@mailbase.ac.uk
	we2hyp-l	National Society of Black Engineers	listserv@psuvm.psu.edu
Entomology	insect-l	Insect physiology	listserv@vm.tau.ac.il
Environmental Science	esanews	Ecological society of America: society news and business	listserv@umdd.umd.edu
	ae	Alternative energy sources	listserv@sjsuvm1.sjsu.edu
	airpollution-biology	Biological/environmental impact of air pollution	mailbase@mailbase.ac.uk
	aquifer	pollution and groundwater recharge	listserv@ibacsata.bitnet
	bes-ecol-aquatic	Aquatic ecology	mailbase@mailbase.ac.uk

Discipline	List Name	Subject	Listserv address
	bes-ecol-comp	Computers in ecology	mailbase@mailbase.ac.uk
	biosph-l	Biosphere ecology	listerv@listserv.aol.com
	biosph-l	Biosphere discussion list	listserv@ubvm.cc.buffalo.edu
	conbio-l	Alberta Chapter of Conservation Biology	listserv@ucnet.ucalgary.ca
	consbio	Conservation biology	listproc@u.washington.edu
	consbio-l	Conservation biology	listproc@cornell.edu
	consgis	Biological conservation and GIS	listserv@uriacc.uri.edu
	conslink	Biological conservation discussion	listserv@sivm.si.edu
	ecdm	Environmentaly concious design and manufacturing	listserv@pdomain.uwindsor.ca
	ecocity	Sustainable urban development	listserv@searn.sunet.se
	ecolog-l	Ecological society of America	listserv@umdd.umd.edu
	ecolview	World views and models for a sustainable society	listserv@listserv.uu.se
	ecosys-l	Ecosystem theory and modelling	listserv@vm.gmd.de
	encon-l	Energy conservation management issues in higher education	listserv@listserv.syr.edu
	energy-l	Energy	listserv@taunivm.tau.ac.il
	envbeh-l	Environment and human behaviour	listserv@polyvm.bitnet
	envinf	Environmental information	listserv@nic.surfnet.nl

Discipline	List Name	Subject	Listserv address
	envinf-l	Environmental information	listserv@nic.surfnet.nl
	envst-l	Environmental studies discussion	listserv@brownvm.brown.edu
	eohsi	Environmental and Occupational Health Sciences Institute	listserv@uci.edu
	et-ann	Ecotechnology for sustainable development	listserv@searn.sunet.se
	glrc	Great Lakes Research Consortium Information	listserv@listserv.syr.edu
	green-travel	Green travel and tourism	majordomo@igc.apc.org
	hydrogen	Hydrogen as an alternative fuel	listserv@uriacc.uri.edu
	mab_ model_ forest	Networking biosphere reserves and model forests	listproc@ucdavis.edu
	mabnet_ america	Electronic networking for MAB biosphere reserves	listproc@udavis,edu
	pacific-biosnet	Natural resources of the Pacific	listproc@listproc.wsu.edu
	raptor-c	Conservation and rehabilitation of raptors	listserv@vm1.spcs.umn.edu
	sfer-l	South Florida Environmental Reader	listserv@ucf1vm.cc.ucf.edu
	sustain	Sustainable development and institutionalism	listserv@irlearn.ucd.ie
	terramon	TERRAMON Long term environmental monitoring in Nfld. and Lab.	listserv@morgan.ucs.mun.ca
	urbnrnet	Urban natural resources	listserv@uga.cc.uga.edu
	usiale-l	International Association of Landscape Ecology	listserv@uriacc.uri.edu
	water-l	Water quality discussion	listproc@listproc.wsu.edu

Discipline	List Name	Subject	Listserv address
Ethics, History and Philosophy	aaasest	Ethical issues in science and technology	listserv@gwuvm.gwu.edu
	aaasmsp	Minority perspectives on ethics in science and technology	listserv@gwuvm.gwu.edu
	biomed-l	Biomedical ethics	listserv@vm1.nodak.edu
	darwin-l	History and theory of the historical sciences	listserv@ukanaix.cc.ukans.edu
	earlyscience-l	History of early science	listserv@listserv.vt.edu
	ecotheol	Ecological theology	mailbase@mailbase.ac.uk
	fastnet	Federation of activists on science and technology	majordomo@igc.apc.org
	gender-set	Research on gender, science, technology and engineering	mailbase@mailbase.ac.uk
	hopos-l	History and Philosophy of Science	listserv@ukcc.uky.edu
	htech-l	History of technology	listserv@sivm.si.edu
	isl-sci	Islam and science	listserv@vtvm1.cc.vt.edu
	medsci-l	Medieval Science Discussion	listserv@brownvm.brown.edu
	medtextl	Medieval texts — philology, codicology and technology	listserv@postoffice.cso.uiuc.edu
	phtech-l	Philosophy and technology	listserv@psuvm.psu.edu
	pol-sci-tech	Democratic politics of science and technology	majordomo@igc.apc.org
	satsunet	Social and economic studies of contemporary science and technology	mailbase@mailbase.ac.uk

Discipline	List Name	Subject	Listserv address
	scichr-list	Science and christianity discussion list	majordomo@eskimo.com
	science-as-culture	Science as culture	listserv@maelstrom.stjohns.edu
	scifraud	Discussion of fraud in science	listserv@cnsibm.albany.edu
	scipol-l	Science Policy Discussion Group	listserv@vtvm1.cc.vt.edu
	scisoctheory-l	Society for Social Studies of Science	listproc@cornell.edu
	slmk	Swedish Physicians Against Nuclear Weapons	listserv@sokrates.mip.ki.se
	smetdial	Science/Math/ Engineering/Technology Dialog	listserv@vm.temple.edu
	stpp	Science, technology and society discussion	listserv@gwuvm.gwu.edu
	stppnews	Science technology and public policy news	listserv@gwuvm.gwu.edu
Evolution	evo-devo	Development and evolution	listproc@lists.colorado.edu
	evolutionary-computing	Analogues of natural selection	mailbase@mailbase.ac.uk
	humevo-l	Human evolutionary research	listserv@freya.cc.pdx.edu
	sociobio	Evolution of social behaviour	listserv@maelstrom.stjohns.edu
Food Science	food-net	Food safety and nutrition	listserv@vm1.spcs.umn.edu
General	acadlabs	Academic labs	listserv@yorku.ca
	coalsglp	COALS good laboratory practices	listserv@tamvm1.tamu.edu

Discipline	List Name	Subject	Listserv address
	devel-l	Technology transfer in international development	listserv@auvm.american.edu
	dost	Turkish scientists discussion	listserv@vm3090.ege.edu.tr
	efos	Discussion forum for Association for Foundation of Science	listserv@plearn.edu.pl
	ensafety	Moderated version of safety@uvmvm.uvm.edu	listserv@peach.ease.lsoft.com
	eqrn-l	Empowering qualitative research	listserv@vm.cc.purdue.edu
	familysci	Family science	listserv@ukcc.uky.edu
	ibrtlist	Institute for Biological Research and Technology List	listserv@univscvm.csd.scarolina.edu
	intdev-l	International development	listserv@uriacc.uri.edu
	inventors	Inventors and the inventing process	listserv@home.ease.lsoft.com
	irsainfo	Irish Research Scientists Association	listserv@irlearn.ucd.ie
	mfm-forum	Magnetic force microscopy	mailbase@mailbase.ac.uk
	mn-wise	Women in science and engineering	listserv@vm1.spcs.umn.edu
	museum-l	Museum curatorial discussion	listserv@unvma.unm.edu
	natosci	Information on NATO science and environment programmes	listserv@cc1.kuleuven.ac.be
	nukop-scitech	UK government science and technology publications	mailbase@mailbase.ac.uk

Discipline	List Name	Subject	Listserv address
	pcst-l	International Network on Public communication of Science and Technology	listproc@cornell.edu
	qualrs	Qualitative research for the human sciences	listserv@uga.cc.uga.edu
	rat-talk	Research animals topics	listserv@nic.surfnet.nl
	regsc-l	Regional sciences exchange	listserv@wvnvm.wvnet.edu
	sas-l	SAS discussion	listserv@uga.cc.uga.edu
	saspac-l	SASS public access consortium	saspac-l@umslvma.umsl.edu
	science-structure	Structure in natural phenomena	mailbase@mailbase.ac.uk
	scifaq-l	Science frequently asked questions	listserv@yalevm.cis.yale.edu
	scinews	News Service Scientific News Releases	listserv@vm.cc.purdue.edu
	spssx-l	SPSS discussion	listserv@uga.cc.uga.edu
	stscan-l	Canadian forum for science and technology	listserv@yorku.ca
	sudan-t	Sudan technology	listserv@emuvm1.cc.emory.edu
	uk-research-methodology	Methodologies of UK manufacturing and operations management research	mailbase@mailbase.ac.uk
	wisenet	Women in Science and Engineering Network	listserv@uicvm.cc.uic.edu
	zst	Zambia science and technology	listproc@msstate.edu
Genetics	aqua-genetics	Genetics of farmed or exploited aquatic organisms	mailbase@mailbase.ac.uk

Discipline	List Name	Subject	Listserv address
	ecol-genetics	Ecological genetics	mailbase@mailbase.ac.uk
	gen-ned	Human genetics in The Netherlands	listserv@nic.surfnet.nl
	gene-l	Genetics education network	listserv@ksuvm.ksu.edu
	oph-gen	Ophthalmogenetics	listserv@nic.surfnet.nl
	pzandqz	Genetic counselling students	listserv@indycms.iupui.edu
Geoscience	geogable	Geography and disabilities	listserv@ukcc.uky.edu
	astra-ug	ASTRA database project users group	listserv@vm.cnuce.cnr.it
	canspace	Canadian space geodesy forum	listserv@unbvm1.bitnet
	clagnet	Conference of latinamericanist geographers	listserv@listserv.syr.edu
	climlist	Climatology	listserv@psuvm.psu.edu
	coastal-research	Coastal environment	mailbase@mailbase.ac.uk
	coastgis	Coastal GIS distribution list	listserv@irlearn.ucd.ie
	coastnet	Coastal management and resources	listserv@uriacc.uri.edu
	cpgis-l	Chinese professionals GIS list	listserv@ubvm.cc.buffalo.edu
	crit-geog-forum	Critical and radical perspectives in geography	mailbase@mailbase.ac.uk
	eigg-info	Shallow (<500 m) geophysics	mailbase@mailbase.ac.uk
	geo-carbonatites	Ingaceous carbonate rocks	mailbase@mailbase.ac.uk

Discipline	List Name	Subject	Listserv address
	geo-computer-models	Computer modelling in geoscience	mailbase@mailbase.ac.uk
	geo-env	Environmental geology	mailbase@mailbase.ac.uk
	geo-materials	Sample preparation in geology and materials science	mailbase@mailbase.ac.uk
	geo-metamorphism	Metamorphic geology	mailbase@mailbase.ac.uk
	geo-mineralisation	Mineral deposits studies	mailbase@mailbase.ac.uk
	geo-tectonics	Tectonics and related issues	mailbase@mailbase.ac.uk
	geoed-l	geology and earth science discussion forum	listserv@uwf.cc.uwf.edu
	geoged	Geography education list	listserv@ukcc.uky.edu
	geogfem	Feminism in geography	listserv@ukcc.uky.edu
	geoglife	Geography national standards	listserv@tamvm1.tamu.edu
	geograph	Geography discussion	listserv@searn.sunet.se
	geology	Geology discussion	listproc@cc.fc.ul.pt
	geopol	Political geography	listserv@ukcc.uky.edu
	geosci-jobs	Geoscience employment opportunities	listproc@eskimo.com
	gmthelp	Generic mapping tools	listserv@soest.hawaii.edu
	imagrs-l	Image processing and remote sensing	listserv@earn.cvut.cz
	itex	International tundra experiment	listproc@lists.colorado.edu
	leftgeog	Socialist/radical geography	listserv@ukcc.uky.edu

Discipline	List Name	Subject	Listserv address
	maps-l	Maps and air photo systems forum	listserv@uga.cc.uga.edu
	meh2o-l	Middle east water	listserv@taunivm.tau.ac.il
	nlgeo	Newfoundland and Labrador geology	listserv@morgan.ucs.mun.ca
	nlrf	Newfoundland and Labrador Research Forum	listserv@morgan.ucs.mun.ca
	nspt-l	South Pole discussions	listserv@vm1.spcs.umn.edu
	odp-l	Weekly reports and site sumaries from the Ocean Drilling Programme	listserv@tamvm1.tamu.edu
	odpmrcdg	Ocean Drilling Programme micropaleontology reference centres	listserv@sivm.si.edu
	org-geochem	Organic geochemistry	mailbase@mailbase.ac.uk
	polar-l	Arctic and antarctic polar list	listserv@upguelph.ca
	quake-l	Science of earthquakes and related phenomena	listserv@listserv.nodak.edu
	quaternary	Quaternary research	listserv@morgan.ucs.mun.ca
	rocks-and-fossils	Rocks, fossils etc.	majordomo@world.std.com
	seism-l	Seismological data distribution	listserv@bingvmb.cc.binghamton.edu
	seismd-l	Seismological discussion	listserv@bingvmb.cc.binghamton
	twsgis-l	The Wildlife society GIS list	listserv@vm1.nodak.edu
	ucgia-l	Universities Consortium on Geographical information and analysis	listserv@ubvm.cc.bufalo.edu
	urbgeog	Urban geography	listserv@listserv.arizona.edu
	volcano	Volcanology	listserv@asuvm.inre.asu.edu

Discipline	List Name	Subject	Listserv address
Marine Science	aosn-l	Autonomous ocean sampling network	listserv@mitvma.mit.edu
	baltsea	Baltic sea science	listserv@ searn.sunet.se
	brine-l	Brine Shrimp discussion	listserv@uga.cc.uga.edu
	cprn	As Above	NA
	crust-l	Crustacean systematics, distribution and biology	listserv@sivm.si.edu
	cturtle	Sea turtle biology and conservation	listserv@nervm.nerdc.ufl.edu
	deepsea	Hydrothermal vent biology	listserv@uvvm.uvic.ca
	diatom-l	Diatom algae research	listserv@iubvm.ucs.indiana.edu
	ecs-pathology	Dissections of whales and dolphins	mailbase@mailbase.ac.uk
	ecs-sperm-whale	Sperm whales	mailbase@mailbase.ac.uk
	fish-ecology	Fisheries ecology	listserv@searn.sunet.se
	fish-telemetry	Telemetry techniques for study of fish	mailbase@mailbase.ac.uk
	fisheries	Fisheries discussion	majordomo@biome.bio.ns.ca
	marine-tech	Marine technology in the UK	mailbase@mailbase.ac.uk
	matbi-l	Marine all taxa biological inventories	listserv@sivm.si.edu
	medsea-l	Marine biology of the Adriatic Sea	listserv@aearn.edvz.univie.ac.at
	mfdr	Methodology in fish disease research	mailbase@mailbase.ac.uk
	mpa-l	Marine protected areas	listserv@morgan.ucs.mun.ca

Discipline	List Name	Subject	Listserv address
	phocoena	Marine mammal necropsy workshops	listserv@sivm.si.edu
	salmonid	Salmonid genome	listserv@morgan.ucs.mun.ca
	shark-l	Sharks and cartilaginous fish	listserv@utcvm.utc.edu
Materials Science	astmsrch	American Society for Testing and Materials search section	listserv@uga.cc.uga.edu
Mathematics	amath-il	Applied mathematics in Israel	listserv@vm.tau.ac.il
	chaoplex	Chaos and complexity theory	listproc@listproc.wsu.edu
	comb-l	Combinatorial mathematics	listserv@cmich.edu
	ewm	European women in mathematics	listserv@vm.cnuce.cnr.it
	ewm-uk	Women in mathematics	mailbase@mailbase.ac.uk
	fam-math	Family mathematics	listserv@uicvm.uic.edu
	graphnet	Graph theory	listserv@vm1.nodak.edu
	ics-l	International Chemometrics Society	listserv@umdd.umd.edu
	mathsoc	Mathematical sociology	listserv@listserv.dartmouth.edu
	mpsych-l	Society for Mathematical Psychology	listserv@brownvm.brown.edu
	n-linear	Chaos, complexity and related theories in relation to social science	listserv@tamvm1.tamu.edu
	ndrg-l	Nonlinear dynamics research group	listserv@wvnvm.wvnet.edu
	nmbrthry	Number theory	listserv@vm1.nodak.edu

Discipline	List Name	Subject	Listserv address
	nyjm-alg	Abstracts of algebra papers in the New York Journal of Mathematics	listserv@cnsibm.albany.edu
	nyjm-an	Abstratcs of analysis papers in the New York Journal of Mathematics	listserv@cnsibm.albany.edu
	nyjm-top	Abstracts of geometry/topology papers in the New York Journal of Mathematics	listserv@cnsibm.albany.edu
	nyjmth-a	Abstracts from the New York Journal of Mathematics	listserv@cnsibm.albany.edu
	quasi-l	Quasiperiodicity — theory and applications	listserv@vm.gmd.de
	samath	Saudi Association for Mathematical Sciences	listserv@saksu00.bitnet
	scattering-theorists	Scattering in continuum mechanics	mailbase@mailbase.ac.uk
	semigroups	Semigroup theory	mailbase@mailbase.ac.uk
	smbnet	Society for Mathematical Biology	listserv@fconvx.ncifcrf.gov
	timeseries	Time series	mailbase@mailbase.ac.uk
Medicine	acb-clin-chem-gen	Clinical biochemistry	mailbase@mailbase.ac.uk
	acta-l	Acta Anatomica	listserv@emuvm1.cc.emory.edu
	aeroso-l	Health effects associated with exposure to aerosols	listserv@nic.surfnet.nl
	aidsbkrv	AIDS Book Review Journal	listserv@uicvm.uic.edu
	amalgam	Mercury poisoning from dental fillings	listserv@listserv.gmd.de
	anest-l	Anesthesiology	listserv@ubvm.cc.buffalo.edu

Discipline	List Name	Subject	Listserv address
	ani-l	Autism Network International	listserv@utkvm1.utk.edu
	autinet	Autism	autinet-request@iol.ie
	autism	Autism and developmental disabilities	listserv@maelstrom.stjohns.edu
	bacchus	Boost Alcohol Conciousness Concerning Health of US Students	listserv@ricevm1.rice.edu
	backmed	Medical back issues duplicates and exchange list	listserv@sun.readmore.com
	braintmr	Brain tumor research/support	listserv@mitvma.mit.edu
	breast-cancer	Breast cancer discussion	listserv@morgan.ucs.mun.ca
	c-palsy	Cerebral palsy	listserv@maelstrom.stjohns.edu
	cancer-l	Cancer	listserv@wvnvm.wvnet.edu
	canchid	Canadian network on Health in International Development	listserv@yorku.ca
	candid-dementia	Care of patients with young-onset dementia	mailbase@mailbase.ac.uk
	carepl-l	Nursing care plans	listserv@ubvm.cc.buffalo.edu
	cel-kids	Celiac/coeliac wheat/gluten-free children	listserv@maelstrom.stjohns.edu
	celiac	Celiac/coeliac wheat/gluten-free	listserv@maelstrom.stjohns.edu
	cfs-l	Chronic fatigue syndrome	listserv@list.nih.gov
	cfs-med	Chronic fatigue syndrome medical list	listserv@nihlist.bitnet
	cfs-news	Chronic fatigue syndrome newsletter	listserv@list.nih.gov

Discipline	List Name	Subject	Listserv address
	cfs-wire	Chronic fatigue syndrome newswire	listserv@maelstrom.stjohns.edu
	child-pharm	Child-Pharm	listserv@maelstrom.stjohns.edu
	clin_neuro-physiol	Neurophysiology	listserv@listserv.umu.se
	clinical_trials	Profession information about clinical trials	majordomo@world.std.com
	cnmt-net	Nuclear medicine technology	listproc@listproc.gsu.edu
	cns-l	Clinical Nurse Specialists list	listserv@listserv.utoronto.ca
	colon	Colon cancer discussion amongst patients, their families and caregivers	listserv@maelstrom.stjohns.edu
	compmed	Comparative medicine, laboratory animals and related subjects	listserv@wuvmd.wustl.edu
	compumed	Role of computers in medicine	listserv@maelstrom.stjohns.edu
	cornea-list	Corneas	listserv@listserv.lsumc.edu
	crc-list	Clinical research co-ordinators	listserv@vm1.spcs.umn.edu
	cshcn-l	Children with special health care needs	listserv@nervm.nerdc.ufl.edu
	cshcs-l	Center for study of health, culture and society	listserv@emuvm1.cc.emory.edu
	cystic-l	Cystic fibrosis support	listserv@yalevm.cis.yale.edu
	d-perio	NIDR periodontal diseases discussion	listserv@list.nih.gov
	d-saliva	Salivary research	listserv@list.nih.gov
	deaf-l	Deafness	listserv@siucvmb.bitnet

Discipline	List Name	Subject	Listserv address
	deaf-mag	Deaf magazine	listserv@listserv.clark.net
	deafblnd	Deaf-blindness	listserv@tr.wosc.osshe.edu
	dental-health	Dental and oral public health	mailbase@mailbase.ac.uk
	dental-public-health	Dental public health	majordomo@list.pitt.edu
	dentst-l	Dental students discussions	listserv@vm.temple.edu
	derm-l	Dermatology	listserv@yalevm.cis.yale.edu
	diabetes	International research project on diabetes.	listserv@irlearn.ucd.ie
	diabetic-max	Development of diabetes related computer-mediated communication projects	majordomo@rachael.franken.de
	dis-sprt	Disability support of families	listserv@maelstrom.stjohns.edu
	dissoc	Dossociative disorders	listserv@maelstrom.stjohns.edu
	down-syn	Down syndrome	listserv@vm1.nodak.edu
	eat-dis	Eating disorders	listserv@maelstrom.stjohns.edu
	eczema	Eczema	listserv@maelstrom.stjohns.edu
	emflds-l	Electromagnetics in medicine, science and communications	listserv@ubvm.cc.buffalo.edu
	eopc-l	Eye on primary care	listserv@ubvm.cc.buffalo.edu
	epiworld	Computer applications in epidemiology	listserv@univscvm.csd.scarolina.edu
	ess-news	Shock, tauma, sepsis and clinical care	listserv@ls.univie.ac.at
	estiv-l	European Society of Toxicology *in Vitro*	listserv@nic.surfnet.nl

Discipline	List Name	Subject	Listserv address
	eyemov-l	Eye movement	listserv@listserv.spc.edu
	family-l	Academic family medicine discussion	listserv@mizzou1.missouri.edu
	fet-net	Research in fetal and perinatal physiology	listserv@.nic.surfnet.nl
	fhmpp-l	Fetal heart monitoring	listserv@ubvm.cc.buffalo.edu
	f-hypdrr	Familial hypophosphatemia vitamin D resistant rickets	listser@maelstrom.stjohns.edu
	fibrom-l	Fibromyalgia/Fibrositis	listserv@vmd.cso.uiuc.edu
	forens-l	Forensic medicine and sciences	mailserv@acc.fau.edu
	fot-group	Forced oscillation technique	mailbase@mailbase.ac.uk
	g.crcsig	Co-ordinating clinical research	g.crcsig-request@milwaukee. va.gov
	gap	AIDS awareness and prevention	listserv@listserv.syr.edu
	gerinet	Geriatric health care	listserv@ubvm.cc.buffalo.edu
	glb-hlt	Global forum on medical education and practice	uicvm.uic.edu
	globalrn	Nurses and other healthcare professionals interested in culture and health	listserv@itssrv1.ucsf.edu
	gnet-l	US Department of HHS: Grants information and updates	listserv@list.nih.gov
	gprsrch	Research in general practice	listserv@cc1.kuleuven.ac.be
	gramic-l	Gramicidin	listserv@brownvm.brown.edu
	healr-l	Collaborative research in the health sciences	listproc@listproc.gsu.edu
	health-l	International discussion on health research	listserv@irlearn.ucd.ie

Discipline	List Name	Subject	Listserv address
	health-promotion	Health promotion and disease prevention researchers	listserv@sokrates.mip.ki.se
	healthre	Health care reform	listserv@ukcc.uky.edu
	hecsa	Health sciences communication	listserv@caligari.dartmouth.edu
	hem-onc	Hematologic malignancies	listserv@maelstrom.stjohns.edu
	hepv-l	Hepatitis support	listserv@maelstrom.stjohns.edu
	hiunews	Health Intelligence Unit news	listserv@yorku.ca
	hlt-net	Initiatives in innovation in health professions	listserv@nic.surfnet.nl
	hospic-l	Hospice care	listserv@ubvm.cc.buffalo.edu
	hunt-dis	Huntingdon's Disease support	listserv@maelstrom.stjohns.edu
	hyceph-l	Hydrocephalus support	listserv@listserv.utoronto.ca
	hypbar-l	Hyperbaric and diving medicine list	listserv@technion.technion.ac.il
	hypnosis	Experimental and clinical hypnosis forum	listserv@maelstrom.stjohns.edu
	immnet-l	Medical immunisation tracking systems	listserv@listserv.dartmouth.edu
	inborn-errors	Inherited metabolic diseases	mailbase@mailbase.ac.uk
	incont-l	Incontinence support	listserv@maine.maine.edu
	jmedclub	Medical journal discussion club	listserv@brownvm.brown.edu
	lactnet	Lactation	listserv@library.ummed.edu
	lasmed-l	Laser medicine	listserv@taunivm.tau.ac.il

Discipline	List Name	Subject	Listserv address
	lowvis	Low vision	listserv@maelstrom.stjohns.edu
	mammo-analysis	Analysis of mammograms	mailbase@mailbase.ac.uk
	medconsu	Medical consumerism	listserv@msu.edu
	medforum	Medical students' forum	listserv@listserv.arizona.edu
	medlab-l	Medical laboratory professionals	listserv@ubvm.cc.buffalo.edu
	medlib-l	Medical libraries discussion	listseerv@ubvm.cc.buffalo.edu
	mednets	Medical communications networks	listserv@ndsuvm1.bitnet
	mednews	Health info-com network newsletter	listserv@asuvm.inre.asu.edu
	mednform	Health informatics	listserv@maelstrom.stjohns.edu
	medphy-l	Medcial physics information services	listserv@vm.akh-wien.ac.at
	medphys	Medical physics	listserv@cms.cc.wayne.edu
	medstu-l	Medical students list	listserv@unmvma.unm.edu
	medsup-l	Medical support list	listserv@@yalevm.cis.yale.edu
	mol-diversity	Molecular diversity for basic research and drug discovery	listserv@listserv.arizona.edu
	motordev	Human motor skill development	listserv@umdd.umd.edu
	mslist-l	Multiple sclerosis discussion support	listserv@technion.technion.ac.il
	mxdiag-l	Clinical molecular diagnostics	listserv@health.state.ny.us
	nephro-rn	Nephrology and transplantation	majordomo@majordomo.srv.ualberta.ca

Discipline	List Name	Subject	Listserv address
	neurl	Neuroscience strategic planning	listserv@uicvm.uic.edu
	neuromus	Neuromuscular research	listserv@maelstrom.stjohns.edu
	nhsc	National Health Service Corps	listserv@ukcc.uky.edu
	nhscwg	National Health Service Corps working group	listserv@ukcc.uky.edu
	nihgde-l	NIH gide to grants and contracts	listserv@list.nih.gov
	nurinfo	Nursing informatics	listserv@listserv.syr.edu
	nurse-rogers	Martha Rogers nursing system	mailbase@mailbase.ac.uk
	nursenet	Nursing	listserv@listserv.utoronto.ca
	nurseres	Nurse researchers	listserv@kentvm.kent.edu
	nutepi	Nutritional epidemiology	listserv@tubvm.cs.tu-berlin.de
	oandp-l	Orthotics and prosthetics	listserv@nervm.nerdc.ufl.edu
	occupational-therapy	Occupational therapy	mailbase@mailbase.ac.uk
	omeract	Outcome measures in rheumatoid arthritis clinical trials	listserv@nic.surfnet.nl
	ophthal	Clinical ophthalmology	listserv@ubvm.cc.buffalo.edu
	optometry	Optometry	optometry-request@mailbase.ac.uk
	orem-l	Orem nursing theory	mailserv@acc.fau.edu
	ovarian	Ovarian problems	listserv@maelstrom.stjohns.edu
	p-atitis	Prostatitis	listserv@maelstrom.stjohns.edu
	panet-l	Medical education and health information	listserv@yalevm.cis.yale.edu
	parkinsn	Parkinson's disease support network	listserv@listserv.utoronto.ca

Discipline	List Name	Subject	Listserv address
	path-l	Pathology	listserv@postoffice.cso.emory.edu
	patho-l	Pathology	listserv@emuvm1.cc.emory.edu
	pbpk-l	Physiologically based pharmokinetics simulation modelling	listserv@listserv.navy.al.wpafb.af.mil
	ped-em-l	Paediatric emergency medicine	listserv@brownvm.brown.edu
	pedsleep	Paediatric sleep	listserv@vm.tau.ac.il
	phnutr-l	Public health nutrition	listproc@u.washington.edu
	php-l	Physicians' home page	lisyserv@archangel.silverplatter.com
	physio	Physiotherapy	mailbase@mailbase.ac.uk
	pnatalrn	Perinatal nursing: practice, education and research	listserv@ubvm.cc.buffalo.edu
	pni	Psychoneuroimmunology list	lstsrv@ccat.sas.upenn.edu
	polio	Polio and post polio syndrome	listserv@maelstrom.stjohns.edu
	pollen-sweden	Pollen reports disemination	mailserv@nrm.se
	prevmed	Preventative medicine	listserv@cmsa.berkely.edu
	prostate	Prostate problems	listserv@maelstrom.stjohns.edu
	prozac	Prozac discussion	listserv@maelstrom.stjohns.edu
	psychiatric-nursing	Psychiatric nursing R&D	mailbase@mailbase.ac.uk
	psynurse	Psychiatric nurses	listserv@maelstrom.stjohns.edu
	ptgeri	Physical therapy / geriatrics	listserv@indycms.iupui.edu
	public-health	Epidemiology and public health	mailbase@mailbase.ac.uk

Discipline	List Name	Subject	Listserv address
	radio-biology	Radiobiology	mailbase@mailbase.ac.uk
	rare-dis	Rare diseases	listserv@maelstrom.stjohns.edu
	rc_world	Respiratory care professionals forum	listserv@indycms.iupui.edu
	rceduc-l	Respiratory therapy education and research	listserv@mizzou1.missouri.edu
	rehab-ru	Physical medicine and rehabilitation in rural/community settings	listserv@ukcc.uky.edu
	ret-pig	Retinitis pigmentosa	listserv@nic.surfnet.nl
	rnphdc-l	Nursing doctoral student forum	listserv@mizou1.missouri.edu
	rplist	Retinal degeneration	listserv@maelstrom.stjohns.edu
	rsi-east	Repetitive strain injury	listserv@maelstrom.stjohns.edu
	rubber	Rubber and latex allergies	listserv@tamu.edu
	safety	Environmental, health and safety issues on campuses	listserv@uvmvm.uvm.edu
	sam-l	Society for Adolescent Medicine	listserv@uconnvm.uconn.edu
	schlrn-l	School nurse network	listserv@ubvm.cc.buffalo.edu
	scr-l	Study of cognitive rehabilitation, traumatic brain injury	listserv@mizzou1.missouri.edu
	seronegative-arthritis	Seronegative arthritis research	mailbase@mailbase.ac.uk
	smcdcme	Continuing medical education	listserv@cms.cc.wayne.edu
	smdm-l	Medical decision-making	listserv@listserv.dartmouth.edu

Discipline	List Name	Subject	Listserv address
	sms-snug	Technical, Operational and business issues involved in the use of SMS products	listserv@gibbs.oit.unc.edu
	snurse-l	International student nursing list	listserv@ubvm.cc.buffalo.edu
	sorehand	Carpal tunnel syndrome, tendonitis etc.	listserv@itssrv1.ucsf.edu
	sport-med	Sports medicine	mailbase@mailbase.ac.uk
	stroke-l	Stroke discussion	listserv@ukcc.uky.edu
	stutt-l	Stuttering: research and clinical practice	listserv@vm.temple.edu
	stutt-x	Stuttering — communication disorders	listserv@asuvm.inre.asu.edu
	surginet	General surgery	listserv@listserv.utoronto.ca
	tbi-prof	Traumatic brain injury professionals	listserv@maelstrom.stjohns.edu
	trauma-l	Trauma and emergency surgery	listserv@listserv.lsumc.edu
	tread	Therapists in recovery from addiction	listserv@maelstrom.stjohns.edu
	tropmed	Tropical medicine	Mail lanfran@yorku.ca to subscribe
	uahealthnet	Arizona Health Center news releases	listserv@listserv.arizona.edu
	update	Information exchange between professionals in the Australian alcohol and other drugs field	majordomo@sydney3.world.net
	vitiligo	Vitiligo support and information	listserv@maelstrom.stjohns.edu
	voceval	Vocational evaluation for rehabilitation	listserv@maelstrom.stjohns.edu

Discipline	List Name	Subject	Listserv address
	whooral-pilot	WHO oral health group	mailbase@mailbase.ac.uk
	witsendo	Endometriosis treatment and support	listserv@listserv.dartmouth.edu
	yorkuwi	Health communication in the Carribean	listserv@yorku.ca
	eletqm-l	Electrochemistry	listserv@brufu.bitnet
	dr-ed	Medical education research and development	listserv@msu.edu
	ingest	Ingestive behaviour	listserv@cuvmb.columbia.edu
Meteorology	gps-iono	GPS for Ionospheric research	listserv@listserv.unb.ca
	met-jobs	Meteorology job opportunities	listproc@eskimo.com
	wx-tropl	Tropical storm and hurricane WX products	listserv@postoffice.cso.uiuc.edu
Microbiology	cyan-tox	Cyanobacterial toxins	listserv@vm3090.ege.edu.tr
	d-oral-l	Oral microbiology/immunology	listserv@list.nih.gov
	lactacid	Biology and uses of lactic acid bacteria	listserv@searn.sunet.se
	mycob-research	Mycobacterial disease research	mailbase@mailbase.ac.uk
	soil-plant-microbe	Soil-plant-microbe interactions	mailbase@mailbase.ac.uk
	spiroc-l	Spirochete research	listserv@wvnvm.wvnet.edu
	warn-l	Antibiotics resisitance	listserv@earn.cvut.cz
Molecular Biology	automated-dna-sequencing	Automated DNA sequencing	mailbase@mailbase.ac.uk

Discipline	List Name	Subject	Listserv address
	chr-x	Molecular genetics of the human X-chromosome and X-linked diseases	listserv@nic.surfnet.nl
	cspmb	Canadian society of plant molecular biology	listserv@pnfi.forestry.ca
	filarial-genome	Filarial nematode genome research	mailbase@mailbase.ac.uk
	mhg	Distribution of bluesheet for molecular and human genetics	listserv@listserv.bcm.tmc.edu
	virolo-l	Molecular virology	listserv@ukcc.uky.edu
	hum-molg	Human molecular genetics	listserv@nic.surfnet.nl
	molecular-cell-speak	Signalling mechanisms within and between between cells	mailbase@mailbase.ac.uk
Neurology	alms-nn	Neural networks	listserv@ua1vm.ua.edu
	ecovis-l	Trends in the ecology of vision	listserv@yalevm.cis.yale.edu
	inns-l	International Neural Network Society	listserv@umdd.umd.edu
Pharmacology	pharmacology-comms	Pharmacology	mailbase@mailbase.ac.uk
Physics	aerosp-l	Aeronautics and aerospace history	listserv@sivm.si.edu
	apasln	APA science leaders network	listserv@gwuvm.gwu.edu
	asipp-l	Chinese plasma physics forum	listserv@ulkyvm.louisville.edu
	cfd	Computational fluid dynamics	listserv@ukcc.uky.edu

Discipline	List Name	Subject	Listserv address
	cfd-l	Discussions of computational fluid dynamics	listserv@ukcc.uky.edu
	chiral-l	Chiral and bi(an) isotrophic microwave materials	listserv@vm.gmd.de
	dasp-l	Digital Acoustic Signal processing	listserv@earn.cvut.cz
	fusion	Redistribution of sci. physics.fusion newsgroup	listserv@vm1.nodak.edu
	holograf	NML holographic storage review	listserv@www.nml.org
	iop-qeg	UK quantum electronics	mailbase@mailbase.ac.uk
	iron-project	Electron excitation cross-sections and rates of astrophysical and technological importance	iron-project-request@mailbase.ac.uk
	maccrug	Opto-electronic recordings	listserv@listserv.umu.se
	marnos-l	Nuclear reactor noise studies	listserv@nic.surfnet.nl
	materi-l	Material science	listserv@vm.tau.ac.il
	meteoptic	Meteorological (atmospheric) optics	listserv@fiport.funet.fi
	micro-el	Microelectronics in Israel	listserv@vm.tau.ac.il
	molecular-dynamics-news	News in molecular dynamics	mailbase@mailbase.ac.uk
	mrswomen	Women in materials science related fields	listserv@cmsa.berkely.edu
	numap	Nuclear utility materials and procurement	listserv@peach.ease.lsoft.com
	optics-l	Optical science in Israel	listserv@vm.tau.ac.il

Discipline	List Name	Subject	Listserv address
	packrnd	Packaging research and development	listserv@vm1.nodak.edu
	phys-stu	Physics student discussion	listserv@uwf.cc.uwf.edu
	physic-l	Physics in Israel	listserv@vm.tau.ac.il
	physics	Physics	listserv@ubvm.cc.buffalo.edu
	physics	Physics	listserv@vm.marist.edu
	polymerp	Polymer physics	listserv@nic.surfnet.nl
	pps	Physics of phase separation	listserv@kentvm.kent.edu
	radsci-l	Radiological science	listserv@western.tec.wi.us
	rbapps	Nuclear industry risk-based technology and applications	listserv@peach,ease.lsoft.com
	saphysia	Saudi physics discussion	listserv@saksu00.bitnet
	semicon-ductors-2-6	II-VI semiconductors	mailbase@mailbase.ac.uk
	sup-cond	Superconductivity	listserv@vm.tau.ac.il
	thermal	Thermal analysis	listserv@msu.edu
	tribo-ej	Tribologie and Rheologie worldwide e-journal	listserv@vm.gmd.de
	applspec	Society for Applied Spectroscopy	listserv@uga.cc.uga.edu
Physiology	maclab	Mac hardware for physiologists	listserv@irlearn.ucd.ie
Plant Science	algae-l	Algae botany	listserv@itlearn.ucd.ie
	brom-l	Bromeliaceae plant family	listserv@ftpt.br

Discipline	List Name	Subject	Listserv address
	cactus-l	Cactus pear	listserv@taiu.edu
	cambium	Cambium modelling group	listserv@listserv.arizona.edu
	cellwall	Plant cell wall research	listserv@vm1.nodak.edu
	cp	Carnivorous plants	listproc@opus.hpl.hp.com
	env-physiol	Plant environmental physiology	mailbase@mailbase.ac.uk
	forest	International students of forest sciences	listproc@u.washington.edu
	gardens	Exchange of information about home gardening	listserv@ukcc.uky.edu
	herb	Medicinal and aromatic plants	listserv@vm3090.ege.edu.tr
	herb-l	Inter-mountain and Pacific Northwest herbarium discussions	listserv@idbsu.idbsu.edu
	hort-l	Virginia Tech Horticulture Department — Monthly Releases	listserv@vtvm1.cc.vt.edu
	lapollen-l	Latin American pollen database	listproc@lists.colorado.edu
	nplc-l	Plant lipids	nplc-l@msu.edu
	peanut-l	Peanut research	listserv@uga.cc.uga.edu
	phy-talk	Phytolith research society	listserv@vm1.spcs.umn.edu
	phytophar-macognosy	Plant based pharmacognosy	mailbase@mailbase.ac.uk
	plant-hormones	Plant hormone research	mailbase@mailbase.ac.uk
	plant-re	Plant sexual reproduction	listserv@nic.surfnet.nl
	plant-taxonomy	Plant taxonomy	mailbase@mailbase.ac.uk

Discipline	List Name	Subject	Listserv address
	plant-tc	Plant tissue culture	listserv@vm1.spcs.umn.edu
	polpal-l	Pollination and palynology	listserv@uoguelph.ca
	root-res	Plant root functioning and architecture	listserv@nic.surfnet.nl
	seed-biology-l	Seed biology research	listproc@cornell.edu
	stressnet	Plant stress	mailbase@mailbase.ac.uk
	treephys	Tree physiology	listserv@vm1.ucc.okstate.edu
	treeseed	Tree seeds	listserv@pnfi.forestry.ca
	aabgacol	Botanical gardens and arboretum staff interested in plant collections	listserv@msu.edu
	hortpgm	Va Tech Horticulture Dept.- Programmes in Consumer Horticulture	hortpgm@vtvm1.cc.vt.edu
Psychology	alcohol-psychol	Psychology of alcohol consumption	mailbase@mailbase.ac.uk
	add_med	Professional discussions on addictive behaviour	listserv@maelstrom.stjohns.edu
	addict-l	Academic and scholarly discussions of addiction related topics	listserv@kentvm.kent.edu
	apasd-l	APA Research psychology network	listserv@vtvm1.cc.vt.edu
	aspsych	Applied social psychology	listserv@gwuvm.gwu.edu
	assess-p	Psychological assessment and psychometrics	listserv@maelstrom.stjohns.edu
	behav-an	behavioral analysis	listserv@vm1.nodak.edu
	behavior	Behavioral and emotional disorders in children	listserv@asuvm.inre.asu.edu

Discipline	List Name	Subject	Listserv address
	borderpd	Borderline personality disorders support	listserv@maelstrom.stjohns.edu
	brit-l	Behavioral research in transplantation	listserv@ksuvm.ksu.edu
	c-psych	Cross-cultural psychology	listserv@maelstrom.stjohns.edu
	capsych	Child psychology	listserv@maelstrom.stjohns.edu
	cdmajor	Communication disorder	listserv@kentvm.kent.edu
	clinipsy	Clinical psychology	listserv@vm1.nodak.edu
	co-occurring-disorders	Co-occurring mental health and substance abuse disorders	listserv@listserv.aol.com
	community-psychology	Discussion of the theory and practice of community psychology	community-psychology @maelstrom.stjohns.edu
	counpsy	Counseling psychology practice and science	listserv@uga.cc.uga.edu
	cpsy-l	Cognitive psychology	listserv@listserv.vt.edu
	downs-research	Social and cognitive functioning of Down's children	mailbase@mailbase.ac.uk
	eawop-l	European Association of Work and Organisational Psychology	listserv@nic.surfnet.nl
	ecopsych	Nature-counselling community connection	listserv@maelstrom.stjohns.edu
	edstyle	Learning styles theory and research	listserv@maelstrom.stjohns.edu
	forenpsy	Forensic psychology	listserv@maelstrom.stjohns.edu
	group-psycho-therapy	Exchange of ideas between group psychotherapy professionals	majordomo@freud.apa.org

Discipline	List Name	Subject	Listserv address
	how-maths-work	Psychology of mathematicians	mailbase@mailbase.ac.uk
	iapsy-l	Interamerican psychologists (SIPNET)	listserv@iubvm.ucs.indiana.edu
	imagery	Concious experience and research in mental imagery	listserv@maelstrom.stjohns.edu
	intpsy-l	International organisations in Psychology	listserv@utepvm.ep.utexas.edu
	ioob-l	Industrial psychology	listserv@uga.cc.uga.edu
	ioobf-l	Industrial psychology forum	listserv@uga.cc.uga.edu
	iopsych	Industrial/Organisational psychology	listserv@gwuvm.gwu.edu
	learning	Learning behaviour models and theory	listserv@maelstrom.stjohns.edu
	madness	User voices in public mental health	listserv@maelstrom.stjohns.edu
	nuvupsy	Against the medical model of behaviour	listserv@maelstrom.stjohns.edu
	ocd-l	Obsessive Compulsive Disorder	listserv@vm.marist.edu
	pain-l	Physical and emotional pain	listserv@maelstrom.stjohns.edu
	pcp	Personal construct psychology	mailbase@mailbase.ac.uk
	pit-d	Psychotherapists in training discussion	listserv@maelstrom.stjohns.edu
	poly-psy	Political science — psychology/psychiatry	listserv@maelstrom.stjohns.edu
	psy-lang	Language and psychology	listserv@maelstrom.stjohns.edu
	psyca	Psychologists practicing in California	listserv@maelstrom.stjohns.edu

Discipline	List Name	Subject	Listserv address
	psyca	Psychologists practicing in California	listserv@maelstrom.stjohns.edu
	psych-couns	Counselling psychology	mailbase@mailbase.ac.uk
	psych-expts	Experiment generator packages in psychology	mailbase@mailbase.ac.uk
	psych-genetic	Psychology of genetic screening and counselling	mailbase@mailbase.ac.uk
	psych-postgrads	Psychology postgraduate research	mailbase@mailbase.ac.uk
	psyche-d	Interdisciplinary research on conciousness	listserv@iris.rfmh.org
	psyche-l	Psyche: an inter-disciplinary electronic journal of research on conciousness	listserv@iris.rfmh.org
	psycho-analytic-studies	Psychoanalysis	listproc@sheffield.ac.uk
	psycho-logical-type	Psychological-type discussion	listserv@listserv.vt.edu
	psylaw-l	Psychology and law	listserv@utepvm.utep.edu
	psyny	Psychological practice in New York	listserv@maelstrom.stjohns.edu
	psysts-l	Psychology statistics	listserv@mizzou1.missouri.edu
	psyusa	USA praticising psychologists	listserv@maelstrom.stjohns.edu
	ptrain	Trainers and trainees in psychiatry forum	listserv@maelstromstjohns.edu
	radical-psychology-network	Radical psychologists	mailbase@mailbase.ac.uk
	research	Psychology of the Internet research and theory	listproc@cmhc.com

Discipline	List Name	Subject	Listserv address
	rorschac	Rorschach discussion and information	listserv@maelstrom.stjohns.edu
	scip-l	Society for Computers in Psychology	mailserv@acc.fau.edu
	yanx-dep	Child and adolescent anxiety and depression	listserv@maelstrom.stjohns.edu
Statistics	genstat	Genstat statistical system	listserv@listserv.rl.ac.uk
	pstat-l	Statistics and programming	listserv@irlearn.ucd.ie
Veterinary Science	avi-l	Association of Veterinary Informatics	listserv@wuvmd.wustl.edu
	pnwvets	Pacific Northwest veterinary discussions	listproc@listproc.wsu.edu
	vetcai-l	Veterinary medicine computer assisted instruments	listserv@ksuvm.ksu.edu
	vetmed-l	Veterinary medicine	listserv@uga.cc.uga.edu
	vetstu-l	Veterinary student discussions	listserv@uga.cc.uga.edu

Newsgroups

Ms Ida Cope, busy science technician in a large metropolitan secondary school, is always frustrated with one particular science demonstration that Mr. Filbert (Head, Science Department) organises; for reasons unknown to her, the experiment never actually works. Ms Cope knows, just knows in her bones, that there's a very simple explanation for the failure of the demo, and she'd love to see the delight on the pupils' faces if it did work. If only she could take advantage of the opportunity to ask other technicians, she's sure she could provide a superior service to the school. But like her, most of the country's technicians aren't really connected into a useful information exchange system.

Now, over the half-term break, Mr. Filbert has taken the plunge and hooked up the computer in the science department prep room to the Internet, but nobody really knows how to use it yet. Ms Cope has a free period on her hands, when all her preparation is finished, and she stares at the beckoning screen. If only she could wave a magic wand, and contact somebody who had actually performed the demo successfully! It could be an interested teacher, or even an enthusiastic student who'd seen it done. If only Ida knew that she just had to open the newsreader software, check out the really useful newsgroups 'uk.education.misc', or 'k-12.ed.science', and cross-post a brief query there for anyone to see. She might well get a note back from someone in say, Canada, that very day outlining a solution to the problem. For goodness sake — somebody give her a copy of this book!

INTRODUCTION

Whenever you read in the newspapers about somebody having read or seen something outrageous on the Internet, it will have been on Usenet. It is the part of the net where triviality, extremism and abuse are most common and where, yes, you can find pictures of people without their clothes on.

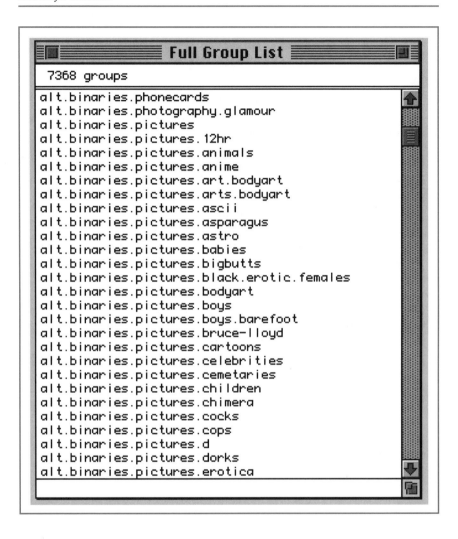

Just in case you've forgotten what that looks like. Yet even on the Usenet Newsgroups, among all the flotsam, jetsam, and big bottoms, there are considerable resources of use to scientists. This chapter will lead you through the Usenet minefield and show you where other scientists can be found.

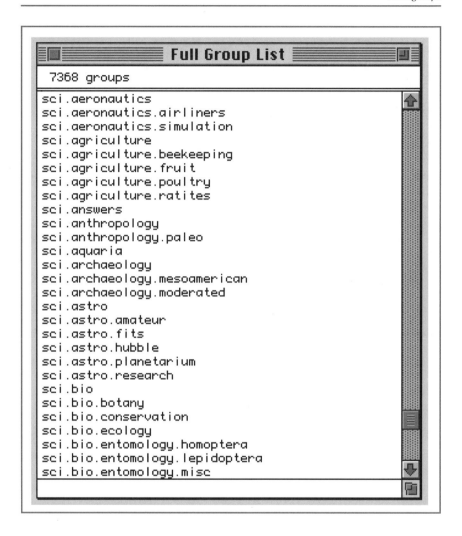

Usenet is a series of global electronic bulletin boards (newsgroups), based on a set of computers which exchange articles to keep a more or less consistent message base on each computer (pedants will tell you that Usenet is the machines that exchange the messages rather than the messages themselves; ignore them, they are just trying to look clever). These articles are grouped into conferences devoted to specific subjects called newsgroups. You read the newsgroup on a subject that interests you and you can comment on the posts already there and add posts of your own. Your additions will be propagated around the world to a potential audience of millions. The detail of the software need not concern us but the gist of it is that the computers on the Internet constantly swap posts to make sure the common message base is up to date.

Even so, messages can sometimes take days to fully propagate themselves and it is not uncommon to be able to read the replies and comments sparked off by a message before the original message itself. A bit of jargon — the comments on a post (and the comments on the comments and so on) are known as a thread, a concept we will return to later. There are no hard rules and no governing body for Usenet. Each sysadmin decides which newsgroups will be carried on the newserver on her site. Despite what you will hear at least some posters say, no one has a right to read any Usenet group they want. It is entirely up to the sysadmins which newsgroups they carry and as long as the people who pay their salaries are happy, that's that. That said, most sysadmins will try to carry as many newsgroups as possible and will add new groups on request or following the announcement of a new newsgroup in the news.announce.newgroup newsgroup, if server space allows.

There are thousands of newsgroups organised in a hierarchical fashion to aid people looking for a particular subject. The main hierarchies, known as 'Big 8', are:

COMP — Computer-related topics; hardware, software etc.

HUMANITIES — Arts faculty type subjects

MISC — Anything that doesn't fit into the other categories or fits into several categories

NEWS — Usenet and news software. Includes key groups where new groups are debated and voted on, and information is made available to newbies.

REC — Recreational activities; sports, hobbies, literature

SCI — groups for various scientific subjects, from the general to the specific

SOC — social issues from politics to emotional support

TALK — "Is this the right room for an argument?"

These categories can sometimes look arbitrary and ramshackle. Partly that's because they are. The hierarchy was drawn up before the Internet became as big as it is today and when the wide range of groups found today was not envisaged. The humanities hierarchy was added as an afterthought some years after the other 7.

From these main categories sub-categories are split off, ending up with groups on sometimes quite specific subjects. The sub-groups are denoted by the full stop. So for example sci.bio.microbiology is a science newsgroup concerned with biology and within biology specifically with microbiology.

The Big 8 are the newsgroups which are available most widely all over the world. Certain rules exist by which new newsgroups can be created and added to the existing Big 8 groups. For those unwilling to go through this process another hierarchy, the alt. hierarchy, exists which doesn't have the same rules but consequently is not carried by as many computers; you pays your money, you takes your choice.

There are other specialist hierarchies, such as bionet which consists of newsgroups of interest to researchers in the biological and medical fields, and k12, of interest to schoolchildren and teachers. We will return to these hierarchies later.

Of the Big 8 newsgroups, the hierarchy of most interest to scientists is sci. The conferences which make up this hierarchy are reviewed at the end of this chapter. We've included screenshots of the sci. hierarchy above in the 'A' section, and below from the tail end:

WHY NOT A MAILING LIST?

As the whole world can read and contribute to the sci newsgroups, their usefulness for professionals is not as great as it could be. However, the newsgroup format still has many advantages over mailing lists. With mailing lists you are presented with a jumble of different messages in your email in-tray that relate to different mailing lists, different topics of discussion and you have to sort through them manually, reading each one. This is obviously time consuming, especially if the list is high volume, as many of the newsgroups are.

With newsgroups, the posts don't arrive in your own computer, you connect to your news server and read them there. You don't need to worry about deleting anything to keep your email in-tray manageable. What's more, they are already arranged into separate topics and posts. What you will see when you read a newsgroup is a list of the titles of recent posts. You can then decide which of those you want to read.

Another important difference is that with a mailing list your posts will be read by an audience of a few hundred people. With a newsgroup the audience is potentially millions. This is both an advantage and a disadvantage. It's an advantage because more people can read and respond to your posts. It's a disadvantage because a great many of them will be non-scientists and some will be cranks. With Usenet, sometimes less really can be more. Like mailing lists, newsgroups have their own internal cultures. There are now so many newsgroups that those on a specific subject tend to be read and contributed to only by those interested in it — the sheer number means it is less likely to be the victim of a 'troll' (a post designed to provoke an argument for the sake of it). Many science newsgroups are so technical that frivolous posts are virtually non-existent.

It is worth pointing out that some newsgroups are "gated" to mailing lists. In other words, messages posted to the newsgroup are also distributed by an email mailing list. This might seem like the worst of both worlds but there are people with no Usenet access who have no choice.

NEWSREADER SOFTWARE

You read newsgroups by connecting up to the news server at your workplace network or Internet provider using newsreader software. At work this will have been supplied to you and it may have been supplied by your Internet provider for a home connection. As with email, there are many different software programmes. The common features are that they will:

1)   Allow you to 'subscribe' to different newsgroups so that when you log in you can look through only those groups and not the whole 10,000 or so. But how to choose which of those thousands are of use? You can't look through them all. Later in this chapter you will be able to go directly to the resources that are relevant to you, using our listing and reviews of science related newsgroups. Here's a list of some of the newsgroups one of our netwatchers software (LW) is 'subscribed' to.

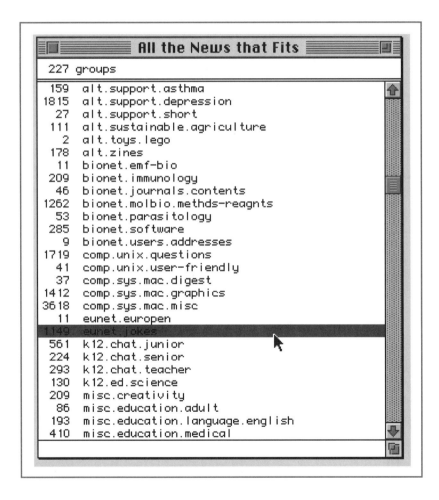

2)   Within the different newsgroups, your newsreader software will first of all show you the titles of the posts, which will themselves be

arranged into discussion threads rather than in simple chronological order. This means that you need only read the posts that interest you.

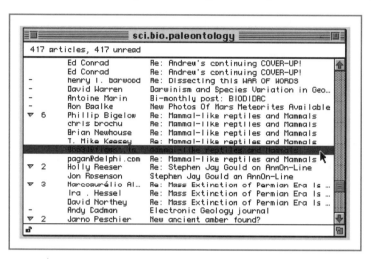

3) You will be able to reply to existing posts in a thread and create new ones. Heres an example of a query, and a helpful reply.

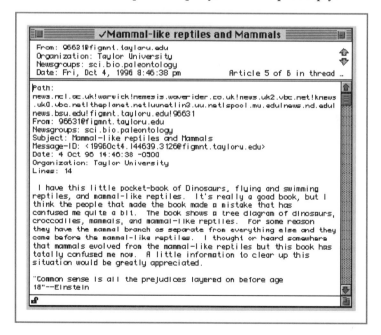

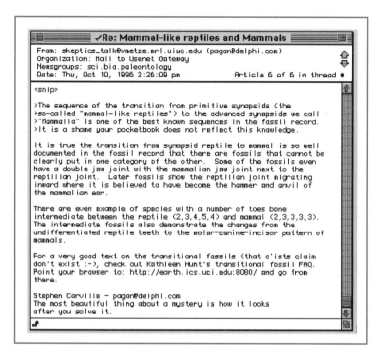

4) Some programmes, for example Free Agent (the chapter dealing with FTP shows how you can obtain this important piece of software for free), have been designed with the home user in mind. They allow you to download the titles of posts in newsgroups you are interested in. You can then mark which messages you would like to read and on your next connection to your ISP, it will download them for you. You can also reply to and make posts off-line and Free Agent will send them to the newsfeed for you the next time you connect. Programmes like this save you money because you can read and reply to messages off-line, saving yourself connection costs.

Essentially that's all there is to newsreader software. The Netscape Navigator software (see the World Wide Web chapter) offers you a slightly different environment in which to read messages:

All you need to do is to be able to work out which of the 10,000 or so newsgroups might be useful to you. It's not always immediately obvious just how to find these useful newsgroups.

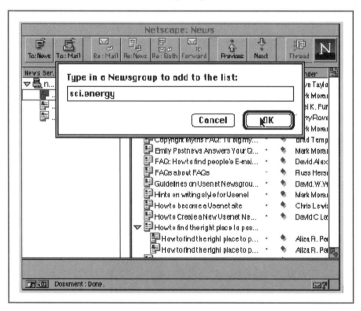

This chapter will help you to do that by reviewing all of the newsgroups concerned with science; you can find their names, and a brief discussion of what goes on in their electronic space.

First however, let's have a look at the 'rules' by which Usenet is run.

NETIQUETTE

As we said before, one of the big advantages of newsgroups is that you can then read all the posts that interest you and ignore the others. For example, you might be interested in the thread with the title 'New sensitive assay for HIV' and so can read those while happily ignoring everything with the title 'HIV is God's punishment to the sodomites!!!!' (yes, such posts do invariably have several exclamation marks and poor grammar).

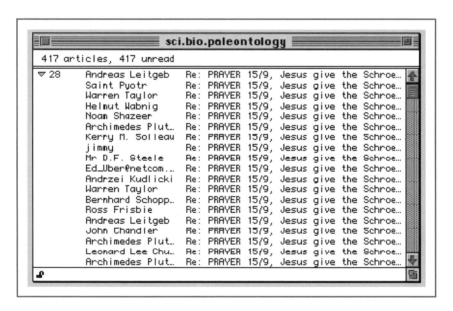

These posts illustrate the big problem with newsgroups, despite their convenience. In the great free speech free-for-all that is Usenet anybody can say anything, anywhere. Yet most people don't — and if you want to be taken seriously neither should you. The fact that most people stick to a few simple rules of netiquette means that the whole thing still runs along relatively smoothly. If you are using a workplace connection, you may find that your access will be withdrawn if enough people complain to your sysadmin about inappropriate or abusive posts from you (not that we think that *you* would ever do such a thing — it's all the other people who might read this book). Like many things concerning the Internet, there is no central authority that makes the rules for Usenet. New groups are created by a mechanism that has a consensus behind it (see below for details) but there's no reason why anyone couldn't decide to short circuit the mechanism. The problem that they would find is that without going through the procedure for naming and approving new groups, the new group wouldn't have the necessary credibility to get propagated among the world's news servers. So, there's no rules but unless you do things in a particular way you won't get very far. We hope that that's clear.

Here are some other non-rules that people generally stick to:

The first non-rule is to post on topic. This means that, as with mailing lists, it's impolite to start up an argument about, for example, why you think Microsoft Windows is the best operating system in a newsgroup dealing with Apple Macintosh software. People who use Apple Macs (poor, deluded souls that they are — KO'D) have already come to an opinion about this, hence their interest in Mac software. Rest assured though, if it's an argument you want there are newsgroups catering for just that — more of them later. So you shouldn't post off-topic. However if you do, you won't be arrested. All that will happen is that a few people will send you mail telling you to stop. If you persist, you'll get a lot more mail telling you to stop. It's the same if you post messages that are abusive: you have every right to do so but people have every right to complain too. As you are reading this book and not *The Internet: All the Porn and Obscenity you can get!* you are probably more concerned with making a good impression with your posts than winding people up for the sake of it.

Imagine though that if instead of posting your witty retort about Macs to the PC conference you posted it to several groups, perhaps even to every single newsgroup on Usenet. Think how many people that would irritate. This practice is known as 'spamming' and is a cardinal sin.

Commercial posts are generally frowned upon and will result in a good quantity of abusive email. This is fair enough, since the last thing you want to read is a load of adverts. There is a separate hierarchy, biz, for commercial postings.

Of course the very best way to make yourself unpopular with millions of total strangers is to combine some of the above e.g. a commercial spam. A firm of lawyers did exactly this and were promptly mail-bombed — sent large volumes of junk data – which led to their internet access being withdrawn.

A good thing to avoid is becoming embroiled in a flame war — an exchange of abusive posts. Just as people are willing to be more open and friendly over the net, they lose their inhibitions about being rude to each other too. If someone replies to one of your posts in a personal, abusive way, react with wit and patience or just drop it. Don't give in to the temptation to reply in kind. Let's face it, you'd be severely embarrassed about having a stand up row with a colleague at a conference or (a more relevant context given some of the characters on Usenet) someone who started haranguing you on the bus. Just think how much more embarrassing it would be to do the same thing in front of millions of people.

Information about Usenet history and advice on netiquette and non-rules is available in the **news.announce.newusers** newsgroup. This newsgroup is made up of a series of regular posts written with the newbie in mind. It is well worth having a look through them. Yes, even if you have read this (or any other) book.

STARTING A NEW NEWSGROUP

So, you've read the newsgroup reviews at the end of the chapter and find to your surprise that there are no newsgroups on marsupials. This omission must be corrected at once — but how? It is a lot more difficult to start a newsgroup than a mailing list — for the very good reason that a 'Big 8' newsgroup will consume resources on computers all over the globe. When you add it all up, that's a lot of resources. Not surprising then that the procedures for setting up a new newsgroup require you to jump through a few hoops first. Of course there are no rules on Usenet but unless you follow the steps described here to the letter, you have no chance of having your group created. Clear? Only the really dedicated computer-freaks can do something like create hundreds of newsgroups spelling out the name of the late Jerry Garcia of the Grateful Dead in number signs:

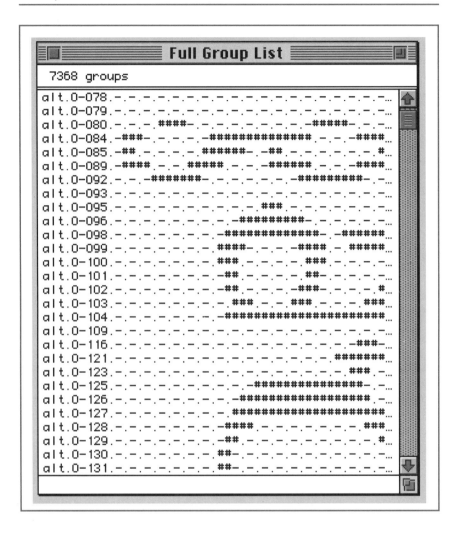

The first thing that you as a scientist need to do is to get the name right. Or, failing that, to get a name which will have as few people vociferously opposing it as possible. Best of all, devise a name that the most annoying people in news.groups will oppose in an abusive fashion, which will guarantee you a huge sympathy vote and victory. Hold on, you say 'vociferous opposition'? 'news.groups' 'vote'? — all I want to do is start up a newsgroup about marsupials, what on earth is going on here? Read on and all will become clear.

The creation of new newsgroups is decided by a vote, following a debate in a newsgroup called news.groups. Anyone, absolutely anyone, can comment on your suggested newsgroup and vote for it or against it. This is a potential minefield but luckily there are friends at hand who can make life simpler. Group-mentors (group-mentors@amdahl.com) are a body of volunteers who will help you through the creation process. Another useful ally is group-advice (group-advice@uunet.uu.net) who will help you to choose an appropriate name — woe betide anyone who proposes sci.marsupials rather than sci.bio.marsupials for example. People take newsgroup names very seriously on Usenet.

The first step is to post a Request for Discussion (RFD) for sci.bio. marsupials to news.announce.newgroups, news.groups and any relevant newsgroups or mailing lists. The RFD consists of the name of the proposed group, and its charter, which sets out what it is for. Discussion takes place in news.groups — theoretically only in news.groups but this has been increasingly ignored with votes being canvassed all over the shop.

After a 30-day discussion period, then a vote can be held on whether the group can be created or not. Votes are handled by a group of people called the Usenet Volunteer Votetakers (UVV), who can be contacted at uvv-contact@uvv.org. There's nothing to say that you can't run your own vote of course. It's just that it wouldn't have the necessary weight to convince news admins that they should allocate resources to your newsgroup. UVV will post a Call for Votes (CFV) to news.groups, which will include the group charter and instructions on how to vote yes or no. The voting period lasts between 21 and 31 days and the closing date will be included in the CFV. After the closing date UVV will count up the votes and publish the result in news.groups, along with the email addresses of everyone who voted, as a safeguard against fraud. The criteria for a result in favour of creation are 100 more yes votes than no votes and at least 2/3 of the total number of votes cast in favour of creation.

If sci.bio.marsupials passes these hurdles then the group will be created, if not then the proposal can't be raised again for another 6 months. 'Created' in this context means that an announcement of the new group is posted to the newsgroup news.announce.newgroups by its moderator. This means that news administrators will add sci.bio.marsupials to the list of newsgroups available at their site – probably.

This might seem like a lot of hard work, not to mention a strange way of going about things. After all, why should non-scientists determine which science newsgroups can be set up? Wouldn't it be great if it was easier to set up science newsgroups and if only scientists voted on them and posted to them? Well, yes it would. The good news is that, if you're a biologist, such a Usenet hierarchy exists: bionet.

BIONET/BIOSCI

Bionet differs in several important respects from other Usenet hierarchies. It was originally planned and funded (by the US National Science Foundation) as a resource for people working in biological research. Consequently it has people who run it and an infrastructure and rules. All of the above is good news for biologists. For one thing, it means that starting a new newsgroup on bionet is much simpler. You have two options. Firstly, you can have any biology research-related group set up as a mailing list for 6 months. During this probationary period the mailing list is maintained and archived for you by BIOSCI and all you need to do is to publicise the list and generate discussion amongst the subscribers. After 6 months, an email vote takes place amongst bionet readers on whether the list deserves to become a full newsgroup. The alternative is to go straight to the vote without the 6 month mailing list period. However if you can make a success of the mailing list then you are virtually guaranteed to win the newsgroup vote.

All in all, bionet newsgroup creation is a much simpler and more civilised process.

In addition, all posts are stored in a searchable archive (http://www.bio.net) which means that they form a permanent useful resource rather than just a transient forum for the exchange of views.

It is a shame that there aren't similar hierarchies for other scientific disciplines.

Nowadays, bionet is no longer funded by the NSF and relies on commercial sponsorship to maintain its wonderful services to biologists. You can help to preserve this invaluable resource by encouraging the companies you deal with to sponsor bionet by taking out an advertisment on the WWW site (http://www.bio.net). If you don't know what a WWW site is then you'll just have to wait until that Chapter.

GETTING STARTED ON USENET

Imagine then, our Professor Realitas. He has his Internet connection and has resolved to have a go and start up his Usenet software, to access the resources he has heard some of his staff and students enthusing about. He clicks on the newsgroup icon on his computer screen.

The first thing he may be presented with is a list of New groups

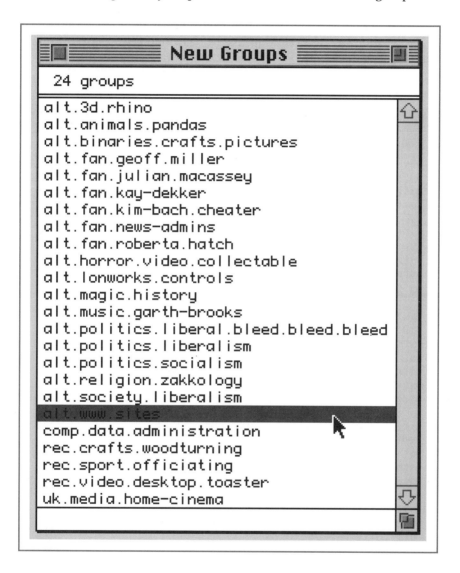

This is not the same as the list of newsgroups — there are well over 10,000 of them, but rather is the list of *new* newsgroups which have been created since his software was last updated. However, as this is the first time he has used his newsreader software, the Prof. is presented with the

entire list of available newsgroups and is invited by the software to subscribe to some. He begins to scroll through the whole group list, enthusiastically at first, to see what science groups he can find. Ten minutes later he has not seen any science groups but he has found newsgroups for every perversion he had heard of, and quite a few he hadn't.

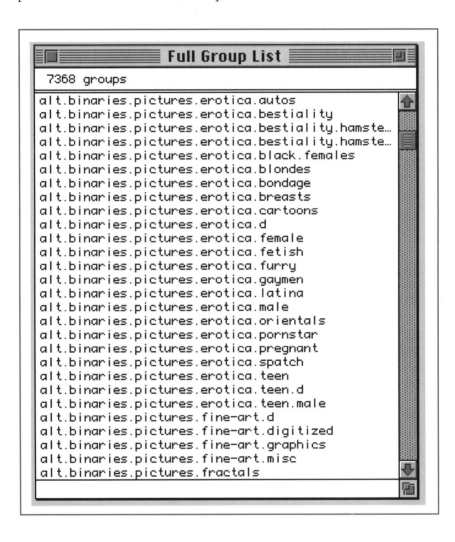

By the time he reaches alt.sex.fetish.star-trek he decides that everything he has heard about the Internet is true and his department's connection should be terminated forthwith.

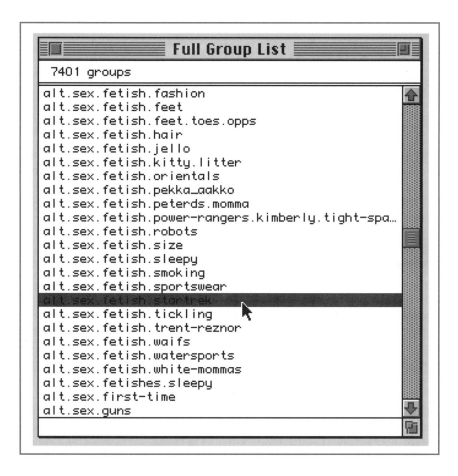

How easily this unpleasantness could be avoided if he had a copy of this book. The next section lists and reviews the Usenet conferences of professional interest to scientists.

Newsgroup Reviews

THE MISC HIERARCHY

The only MISC newsgroup of direct relevance to science is

misc.education.science

This is a low/medium volume group dealing with science education. Examples: "How should evolution be taught?", "Educational WWW sites"

THE SCI HIERARCHY

The sci hierarchy contains some newsgroups which have only a tenuous connection with science and sometimes only a tenuous connection with the real world. That said, most of its newsgroups deal with serious scientific subjects. Some are more prone to frivolous posts than others. A general rule of thumb would seem to be that the more technical the subject, the less trouble there is from frivolous posts. Unlike the bionet hierarchy (see later) sci is open to professionals and public alike. The downside of this is that the content is not always of a high scientific level. The upside is that members of the public are listening and asking questions about science. We've all complained despairingly about the public's lack of interest in science at one time or another — now is your chance to do something about it. If, for example, you've complained that most people don't know the difference between bacteria and viruses, and someone makes a post entitled "Can anyone tell me the difference between bacteria and viruses?" then that's your big chance to do something about it. At the same time you can be participating in an entirely different thread in the same newsgroup about technical aspects of virus detection. Isn't Usenet great?

Here are our quick reviews of the sci newsgroups currently available. By 'volume' we mean the number of posts per day. We have arbitrarily defined low as being less than 5 per day, medium as 5–15 and high as more than 15. Some newsgroups can get more than 100 posts per day — but not in the sci hierarchy — yet. We also mention 'signal to noise ratios', by which we mean the proportion of on-topic posts to trivia, cranks, spam and other annoyances.

sci.aeronautics

This is a low volume, moderated group on scientific and technical aspects of aeronautics. Sample posts are "airflow sensors" and "Looking for optimal design for propellers for a wind generator".

sci.aeronautics.airliners

A low volume, moderated group on scientific and technical aspects of airliners. Includes a fair number of posts on not strictly technical aspects of airliners. Yes, I suppose it's fair to say that there is a plane-spotting contingent. Examples: "Report of 757 crash", "Weight of cabin air at altitude".

sci.aeronautics.simulation

A low volume, moderated group about flight simulation.

sci.agriculture

A medium volume group for farming and agriculture topics. There is some noise — for example one sporadically arising thread concerns hemp production. Its proponents decry the fact that it's illegal and give long lists of the uses of this crop: clothing, rope, fuel, construction material. Why, they'll be smoking it next. Examples: "Effect of boron on crops", "Crop management software — recommendations?".

sci.agriculture.beekeeping

A low to medium group for apiarists. Everything from bee diseases to honey harvesting. Examples: "Test for varroa", "Preferred smoker material and methods".

sci.answers

This moderated newsgroup consists of periodic postings of FAQs for the different sci newsgroups. Not all sci newsgroups have FAQs but you should look in this conference to see if there are any relevant to your interests. It might save you time and embarrassment — you may also find pointers towards other resources for your subject of interest.

sci.anthropology

Medium volume discussions amongst professional and amateur anthropologists. some noise but perhaps less than you'd expect. Examples: "Aborigine art", "Human skull differences".

sci.anthropology.paleo

Low/medium volume discussion of fossils of man and other primates. Mercifully free of Creationist wars. Examples: "Neanderthal man", "Human phylogeny".

sci.aquaria

Medium volume group dealing with scientifically-oriented aspects of aquaria. Not a subject on most science faculty curricula I suspect — this is basically a hobby group. There are also alt.aquaria and rec.aquaria groups making it one of the best-catered for subjects on Usenet. There is an exhaustive FAQ. Examples: "Snails and seaweed", "What type of filter should I use?".

sci.archaeology

A medium volume group on antiquities. This group has a poor signal to noise ratio with a large proportion of posts on theories about the pyramids, Arthurian legends and so on. Not surprisingly, a new, moderated group has been proposed. Examples: "Why Erich Von Daniken was right", "Pyramids and the stars".

sci.archaeology.mesoamerican

A medium volume group on archaeology as relating to the native civilisations of the Americas. A tightly-oriented group with a good signal to noise ratio, despite the lack of moderation. This is at least partly due to the excellent FAQ. Examples: "Mesoamerica/hebrew link?", "Ceremonial sites in Mexico".

sci.astro

A high volume group for those with an interest in astronomy. Posts are on-topic but tend to cater for the serious amateur rather than the professional. Examples "What sort of telescope should I buy?" and "When comet X can be seen".

sci.astro.amateur

Even higher volume group with more of the same. Examples: "Good places to observe west of Boston", "Mirror materials".

sci.astro.fits

Low volume group on Flexible Image Transport Systems.

sci.astro.hubble

Low volume, moderated, group consisting of daily and other reports containing Hubble telescope data.

sci.astro.planetarium

Low volume newsgroup about planetaria. High proportion of cross-posts from the other sci.astro groups. Examples: "Is there still a planetarium in X", "Planetarium Y show".

sci.astro.research

Low volume, moderated group on astronomical research, including job vacancies. Examples: "Shape of Jupiter", "New Space Horizons conference".

sci.bio

Although this group may still be listed in some lists of newsgroups it no longer exists. It has been replaced by sci.bio.misc (see later).

sci.bio.botany

Medium volume group on plant science. A good mixture of professional and amateur botanical posts. Examples: "Info on Bromeliaceaie sought", "Starch metabolism".

sci.conservation

Low volume, moderated group on conservation biology.

sci.bio.ecology

High volume group on ecology research.

sci.bio.entymology.homoptera

Low volume newsgroup about homopterans (sap-sucking insects). Examples: "When do cicadas emerge?", "Viruses spread by aphids".

sci.bio.entymology.lepidoptera

Medium volume newsgroup on butterflies and moths. Has a good mix of professional and amateur contributions. Examples: "Question about metamorphosis", "Artificial diet recipes".

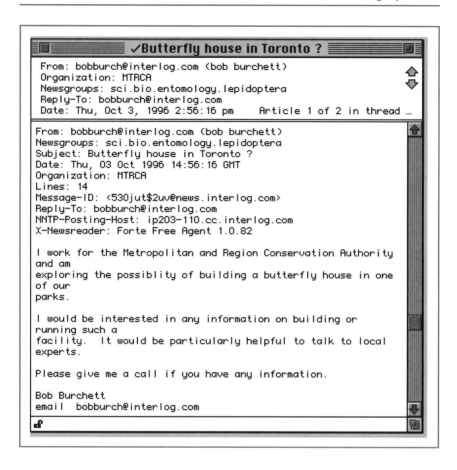

✓Butterfly house in Toronto ?

From: bobburch@interlog.com (bob burchett)
Organization: MTRCA
Newsgroups: sci.bio.entomology.lepidoptera
Reply-To: bobburch@interlog.com
Date: Thu, Oct 3, 1996 2:56:16 pm   Article 1 of 2 in thread ...

From: bobburch@interlog.com (bob burchett)
Newsgroups: sci.bio.entomology.lepidoptera
Subject: Butterfly house in Toronto ?
Date: Thu, 03 Oct 1996 14:56:16 GMT
Organization: MTRCA
Lines: 14
Message-ID: <530jut$2uv@news.interlog.com>
Reply-To: bobburch@interlog.com
NNTP-Posting-Host: ip203-110.cc.interlog.com
X-Newsreader: Forte Free Agent 1.0.82

I work for the Metropolitan and Region Conservation Authority
and am
exploring the possiblity of building a butterfly house in one
of our
parks.

I would be interested in any information on building or
running such a
facility. It would be particularly helpful to talk to local
experts.

Please give me a call if you have any information.

Bob Burchett
email bobburch@interlog.com

sci.bio.entymology.misc

Low-medium volume group covering general discussions about insects.
Examples: "Did flea circuses really exist?", "Forensic entymology".

sci.bio.ethology

Medium volume newsgroup, ostensibly about animal behaviour and
behavioural ecology but about 50% of posts at time of writing were
unrelated trivia. Moderation was on the horizon. Examples: "Parent-off-
spring conflict in social insects", "Predator-livestock conflicts".

sci.bio.evolution

Medium volume, moderated newsgroup with high standard posts on all aspects of evolutionary theory. Newsgroup is moderated to screen out evolution/creationist flame wars and other trivia. Examples: "Co-evolving parasites", "Evolution of altruism".

sci.bio.fisheries

Low volume group on fisheries, fish farming and fishes. Examples: "Thawing bulk fish". "Trout feed".

sci.bio.food-science

Medium volume group on various aspects of food science and technology. Has a comprehensive FAQ that is a good starting point for interested readers. Examples: "processing method for mozzarella", "Use of thickeners".

sci.bio.herp

Medium volume group consisting of professional and amateur discussions on reptile and amphibian biology. Examples: "Frogs in the UK", "Effect of temperature on sex determination in reptiles".

sci.bio.microbiology

Medium volume group on microbiology. Has a good signal to noise ratio, with most posts having a scientific content. However there is a high level of cross-posting with bionet.microbiology. The rationale for the creation of sci.microbiology was that it would be a forum for posts from lay people as well as scientists. Examples: "Pseudomonas denitrification", "Does my dog have more bacteria than me?".

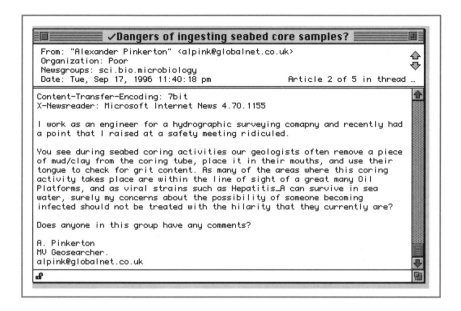

✓Dangers of ingesting seabed core samples?

```
From: "Alexander Pinkerton" <alpink@globalnet.co.uk>
Organization: Poor
Newsgroups: sci.bio.microbiology
Date: Tue, Sep 17, 1996 11:40:18 pm          Article 2 of 5 in thread ...
```

```
Content-Transfer-Encoding: 7bit
X-Newsreader: Microsoft Internet News 4.70.1155

I work as an engineer for a hydrographic surveying comapny and recently had
a point that I raised at a safety meeting ridiculed.

You see during seabed coring activities our geologists often remove a piece
of mud/clay from the coring tube, place it in their mouths, and use their
tongue to check for grit content. As many of the areas where this coring
activity takes place are within the line of sight of a great many Oil
Platforms, and as viral strains such as Hepatitis_A can survive in sea
water, surely my concerns about the possibility of someone becoming
infected should not be treated with the hilarity that they currently are?

Does anyone in this group have any comments?

A. Pinkerton
MV Geosearcher.
alpink@globalnet.co.uk
```

sci.bio.misc

This is a medium volume group on the biological sciences. At the time of writing it was getting increasingly buried under a mound of frivolous posts. Hopefully the circle will turn and it will resume its place as a useful general biology group. Examples: "Is a biology PhD worth it?", "Why are veins blue?".

sci.bio.paleontology

Low volume group on palaeontology, with a good signal to noise ratio. creationism/evolution flame wars are discouraged. "Fossil hunting sites in X", "Help needed to identify fossil from Cretaceous period".

sci.bio.phytopathology

Low volume group dealing with plant disease, pests and pathogens. This group has yet to really take off and is a good illustration of the moderation dilemma: every post is on topic but there's not very many of them. Examples: "Nematode diseases", "PCR for fungal diagnostics".

sci.bio.systematics

Low volume group on systematics and taxonomy. Examples: "Systematics statistics", "Tree of Life WWW page URL".

sci.bio.technology

Medium volume group on biotechnology and other applied biology topics. Has a surprising amount of noise for such a practical subject. Examples: "Patenting", "Bioreactor maintenance".

sci.chem

High volume group on chemistry. Has an excellent signal to noise ratio. A good balance between the professional and the layperson. examples: "Processes for TiO_2 — catalysed destruction of organics", "Assay for Mg^{++} needed".

sci.chem.analytical

Medium volume group on analytical chemistry. Examples: "Thin layer chromatography methods", "UV spectrum for aluminium".

sci.chem.coatings

Low volume newsgroup on paints and other coatings. Examples: "Measurement of colour", "UV resistant pigments".

sci.chem.electrochem

Medium volume group on electrochemistry. A lot of cross-posts with sci.chem.electrochem.battery. Examples: "Nodular deposits of copper", "Redox potentials".

sci.chem.electrochem.battery

Medium volume group about technical and practical aspects of different kinds of battery. Examples: "Charging a nicad battery", "Lithium ion batteries".

sci.chem.labware

Low volume group for discussions of different types and models of laboratory equipment. Mainly taken up with 'for sale' notices for second-hand equipment.

sci.chem.organomet

Low volume group ostensibly about organometallic chemistry but with a proportion of posts about more general aspects of metals e.g. corrosion. Examples: "Titanocenes and N2 fixation", "Monomethylarsonic acid — supplier wanted".

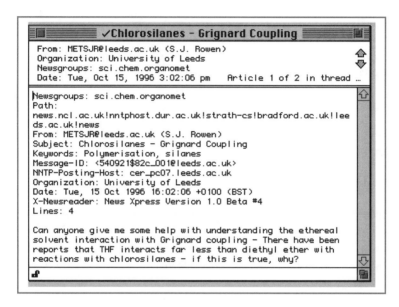

```
✓Chlorosilanes - Grignard Coupling

From: METSJR@leeds.ac.uk (S.J. Rowen)
Organization: University of Leeds
Newsgroups: sci.chem.organomet
Date: Tue, Oct 15, 1996 3:02:06 pm    Article 1 of 2 in thread …

Newsgroups: sci.chem.organomet
Path:
news.ncl.ac.uk!nntphost.dur.ac.uk!strath-cs!bradford.ac.uk!lee
ds.ac.uk!news
From: METSJR@leeds.ac.uk (S.J. Rowen)
Subject: Chlorosilanes - Grignard Coupling
Keywords: Polymerisation, silanes
Message-ID: <540921$82c_001@leeds.ac.uk>
NNTP-Posting-Host: cer_pc07.leeds.ac.uk
Organization: University of Leeds
Date: Tue, 15 Oct 1996 16:02:06 +0100 (BST)
X-Newsreader: News Xpress Version 1.0 Beta #4
Lines: 4

Can anyone give me some help with understanding the ethereal
solvent interaction with Grignard coupling - There have been
reports that THF interacts far less than diethyl ether with
reactions with chlorosilanes - if this is true, why?
```

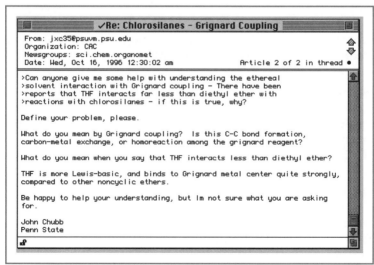

```
✓Re: Chlorosilanes - Grignard Coupling

From: jxc35@psuvm.psu.edu
Organization: CAC
Newsgroups: sci.chem.organomet
Date: Wed, Oct 16, 1996 12:30:02 am    Article 2 of 2 in thread ●

>Can anyone give me some help with understanding the ethereal
>solvent interaction with Grignard coupling - There have been
>reports that THF interacts far less than diethyl ether with
>reactions with chlorosilanes - if this is true, why?

Define your problem, please.

What do you mean by Grignard coupling?  Is this C-C bond formation,
carbon-metal exchange, or homoreaction among the grignard reagent?

What do you mean when you say that THF interacts less than diethyl ether?

THF is more Lewis-basic, and binds to Grignard metal center quite strongly,
compared to other noncyclic ethers.

Be happy to help your understanding, but Im not sure what you are asking
for.

John Chubb
Penn State
```

sci.classics

Medium volume discussions on the study of classical philosophy, history, art and related topics. Not a subject prominent in many science faculties, we suspect. Examples: "Latin for websurfing?" "'Plato' and 'Aristotle' as schools".

sci.cognitive

Medium volume group on perception, memory, judgement and reasoning. Examples: "The physical basis of consciousness", "Vision modelling".

sci.comp-aided

A low volume group on the use of computers as a tool in scientific research.

sci.cryonics

Medium volume group concerning the possibility / practicality of freezing people for later revival. Not part of many science curricula I suspect but evidently there are many people out there who take these things seriously. Examples: "Freezing damage", "Hey has anybody seen that Woody Allen film 'Sleeper'?"

sci.crypt

Medium volume group for discussion of encryption and decryption of data. Examples: "PGP", "Enigma history".

sci.crypt.research

Low volume, moderated group for discussion of cryptography, cryptanalysis and related issues — particularly technical aspects. Example: "Call for papers / programme for conference X".

sci.data.formats

Low volume group on the different formats used for modelling, storage and retrieval of scientific data. Exhaustive FAQ. Examples: "Saving DAT to disk", "NetCDF format".

sci.econ

High volume group on economics. They call it the dismal science. We don't call it a science at all and have only included it for completeness. Examples: "Make Money Real Fast".

sci.econ.research

Even more dismal, though lower volume, than the above.

sci.edu

Medium volume group on the science of education. Examples: "Using the WWW in schools" "Need ideas for physics experiments".

sci.electronics

Obsolete group — now replaced by sci.electronics.misc

sci.electronics.basics

High volume group on basic (household) electronics. Questions from beginners are welcome. Examples: "How to read schematics", "Help needed on circuit design".

sci.electronics.cad

Low/medium volume group on Computer Aided Design software and its use in designing electronic circuits and assemblies. Examples: " Freeware CAD software", "Best printed circuit board editor?".

sci.electronics.components

High volume group on the properties and suppliers of electronic components. Examples: "Need recommendations for capacitor reset switch", "Magnetic field sensors".

sci.electronics.design

High volume group on the design of electronics circuits. Examples: "Impedance matching", "How to measure inductive AC power".

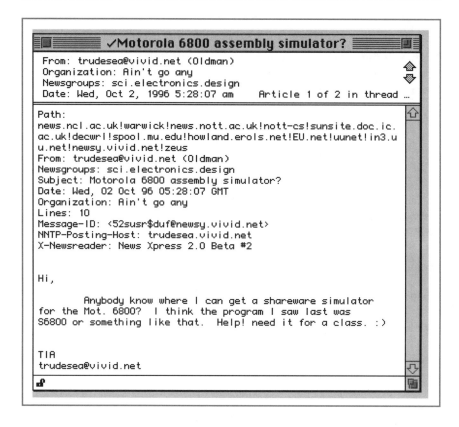

sci.electronics.equipment

Medium volume group on non-domestic electronic equipment. Some noise. 'For sale' messages are discouraged. Examples: "multi-channel analyser wanted- recommendations?", "calibration procedures".

sci.electronics.misc

High volume group on all aspects of electronics — except relative merits of household appliances. Examples: "Hum through my speakers", "International wall-plug standards".

sci.electronic.repair

Medium volume group on repair of electronic equipment. Examples: "Looking for supplier for part X", "CD player has fault Y — any ideas?".

sci.energy

High volume group on energy science and technology. A large proportion of posts with, shall we say, speculative ideas for energy generation. Examples: "Can nuclear waste be used?", "Gas turbine heat transfer".

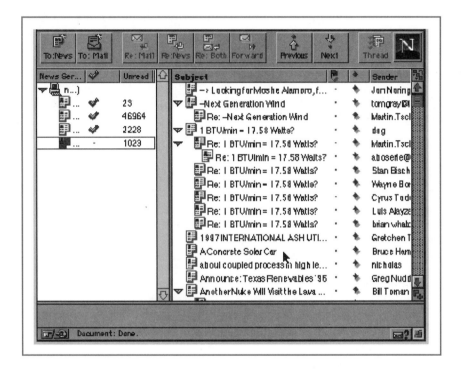

sci.energy.hydrogen

Low volume discussions on the use of hydrogen as an energy source. Examples: "Automotive hydrogen generator", "Extent of use of gas engines".

sci.engr

Medium/high volume group on all aspects of engineering. Examples: "Resistance of wood", "Heat transfer".

sci.engr.biomed

Medium volume group on biomedical engineering. Attracts a lot of noise for some reason. Examples: "Artificial muscle", "Impedance of skin".

sci.engr.chem

Medium volume group on chemical engineering. Shouldn't this be under chemistry you say? Welcome to Usenet. Examples "How to melt magnesium", "Production of styrene".

sci.engr.civil

Medium/high volume group on civil engineering. Examples: "water main leakage detection", Tunnel liner plates".

sci.engr.color

Low volume group on colour use/detection in engineering. Examples: "Colour quantification and detection", "Colour blindness testing".

sci.engr.control

Medium/high volume group on engineering of control systems. Examples: "Detection of cavitation in pumps", "Humidity sensors".

sci.engr.geomechanics

Low volume group on geomechanics. Example: "Stress distribution around cased wellbore".

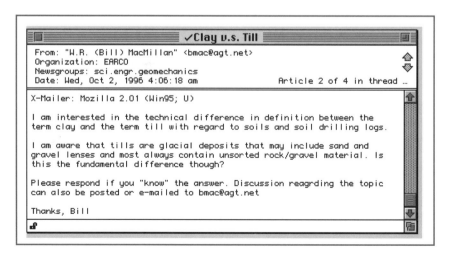

sci.engr.heat-vent-ac

Medium volume group on heating, ventilation, air conditioning systems and refrigeration. Examples: "Screw vs absorption chillers", Central or distributed air filters?".

sci.engr.lighting

Medium/high volume group for lighting, vision, color in architecture, engineering etc. Examples: "Simulation of daylight", "Mercury vapour street lights".

sci.engr.manufacturing

Medium volume group on engineering in manufacturing. Examples: "Alternatives to bar-coding", "Automated lathe machines".

sci.engr.marine.hydrodynamics

Low volume group on marine hydrodynamics. High proportion of cross-posts from other sci.engr groups. Examples: "Yacht simulation", "Boat hull drag".

sci.engr.mech

High volume group on mechanical engineering. Examples: "Strain gauges", " Garbage sorting systems".

sci.engr.metallurgy

Medium volume group on metallurgical engineering. Examples "High strength fatigue wire", "Welding stainless steel".

sci.engr.safety

Low volume group on safety of engineered systems. Examples: "Slip-resistant surfaces", "Under-reporting of incidents".

sci.engr.semi-conductor

Medium volume group on semiconductors processes, materials and physics. Examples "Silicon bulk recombination parameters needed", "Clean room maintenance".

sci.engr.surveying

Low volume group on measurement and mapping of the earth's surface. Examples: "Wetland delineation", "Trig level problem".

sci.engr.television.advanced

Low volume group on HDTV/DATV standards, equipment practices and other matters. Examples: "International workshop on HDTV", "Non-interlaced video".

sci.engr.television.broadcast

Medium volume group on television broadcasting equipment and practices. Examples: "Surround sound", "4:2:2 sampling vs 4:2:0".

sci.environment

High volume group on environment and ecology. Some noise, as might be expected with a subject of general interest. Moderation is being mooted. Examples: "Alternative fuels", "Uptake of heavy metals".

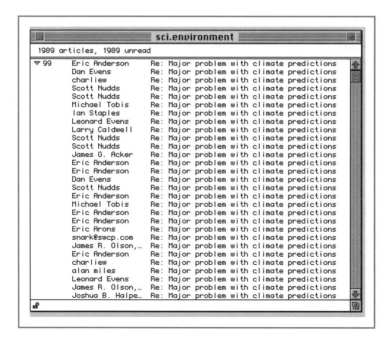

sci.fractals

High volume group on objects of non-integral dimension. Examples: "Practical uses of fractals?", "M-set in polar co-ordinates".

sci.geo.earthquakes

Low/medium volume group on earthquakes and related matters such as volcanic activity. Examples: "Weekly USGS quake report", "Seismographs".

sci.geo.eos

Low volume group on NASA's earth observation system. Comprehensive satellite imaging FAQ. Examples: "Physical oceanography data", "Estimates of Arctic sea melt".

sci.geo.fluids

Low volume group on geophysical fluid dynamics. Many crossposts with other sci.geo groups. Example: "Ocean modelling software wanted".

sci.geo.geology

High volume group on solid earth sciences. Some noise and cross-posting. Examples: "Origin of greenstone belts", "Rock and mineral collecting".

sci.geo.hydrology

Medium volume group on surface and groundwater hydrology. Examples: "Sulphates in rivers", "New model for aquifer characterisation and simulation.

sci.geo.meteorology

High volume group on meteorology. Examples: "Global warming", "How wind speed affects ambient temperature".

sci.geo.oceanography

Low/medium volume group on all aspects of oceanography, oceanology and marine science. Examples: "Measuring salinity", "Gulf of Mexico currents".

sci.geo.petroleum

Medium/high volume group on petroleum and the petroleum industry. Examples: "Advice needed on well-drilling process", "West African oil-fields".

sci.geo.rivers+lakes

Low volume group on the science of rivers and lakes. Examples: "Lake diving", "Use of rainwater tanks for domestic water supply".

sci.geo.satellite-nav

Medium/high volume group on satellite navigation systems, especially GPS. Examples: "GPS navigation formulas", "Recommendations for hand-held GPS?".

sci.image-processing

Medium volume group on scientific image-processing and analysis. Comprehensive satellite imagery FAQ. Good signal to noise ratio. Examples: "Real time image solution for instruments", "Pixel jitter in frame grabbers".

sci.lang

High volume group on natural languages, communication and related matters. Another subject that stretches the definition of science beyond its normal boundaries. Examples: "Pronunciation of Slavic words", "Use of gender in Swedish".

sci.lang.japan

High volume group on the Japanese language. As above. Examples: "How do you say 'call this science?' in Japanese?", "Classical Japanese spelling".

sci.lang.translation

Medium volume group on language translation. Again, does not figure prominently on most science curricula. Examples: "Translation of urdu words requested", "Need help with Arabic boxing terminology".

sci.life-extension

Medium volume group on staying alive longer. Life's too short for this kind of stuff. Examples: "Ozone therapies may prolong life", "Lecithin fuels the brain".

sci.logic

Medium volume group on mathematical, philosophical and computational aspects of logic. Some noise as you might expect, since every crank thinks that their ideas are entirely logical. Examples: "What is a coinductive definition?", "Transfinite induction".

sci.materials

High volume group on materials engineering. Examples: "Materials used in tank armour", "Thermal conductivity of diamond".

sci.materials.ceramics

Low volume groups on ceramic science. Another group that verges towards the hobby end of the spectrum. Examples: "Ceramic pigments", "Coating material for ceramics".

sci.math

High volume group on all aspects of mathematics. Good signal to noise and mix of amateurs and professionals. People actually enjoy maths as a hobby! Examples: "Help needed with algorithm", "A conjecture in algebra theory".

sci.math.numerical-analysis

High volume group on numerical analysis. Examples: "Numerical methods for optimal control", "Numerical differentiation software".

sci.math.research

Medium volume, moderated group on mathematics research. Examples: "Extending group isomorphisms", "Volumes of hyperbolic simplicies and cubes".

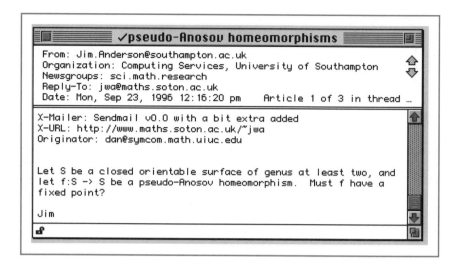

The window titled "✓pseudo-Anosov homeomorphisms" contains:

```
From: Jim.Anderson@southampton.ac.uk
Organization: Computing Services, University of Southampton
Newsgroups: sci.math.research
Reply-To: jwa@maths.soton.ac.uk
Date: Mon, Sep 23, 1996 12:16:20 pm     Article 1 of 3 in thread …

X-Mailer: Sendmail v0.0 with a bit extra added
X-URL: http://www.maths.soton.ac.uk/~jwa
Originator: dan@symcom.math.uiuc.edu


Let S be a closed orientable surface of genus at least two, and
let f:S -> S be a pseudo-Anosov homeomorphism.  Must f have a
fixed point?

Jim
```

sci.math.symbolic

Medium volume group on symbolic algebra. Much of content revolves around software for symbolic algebra. Examples: "Help needed on X software", "Symbolic maths solution to partial differential equations".

sci.mech.fluids

Medium volume group on fluid mechanics. Examples: "Turbulence in fluids", "Two phase fluid flows".

sci.med

High volume group on all aspects of medicine. Expect a fair bit of noise about matters of medical controversy. Examples: "Circumcision is good/bad", "What is gas gangrene?".

sci.med.aids

High volume, moderated group on pathology, biology, treatments and prevention of HIV/AIDS . Examples: "AIDS/HIV clinical trials", "Latest FDA drug approvals".

sci.med.cardiology

Medium/high group on cardiovascular diseases. For a general group there is very little noise — perhaps because, as the major cause of death, it is a little too close to home? Examples: "Exercise physiology and cardiac rehab", "Ventricular tachycardia".

sci.med.dentistry

High volume group on dentistry and related matters. Examples: "Water fluoridation", "Wisdom teeth: yank vs surgery".

sci.med.diseases.als

Low/medium volume group on ALS (atrero-lateral sclerosis). Examples: "Computer systems to aid ALS patients", "Environmental factors in ALS".

sci.med.diseases.cancer

High volume group on diagnosis, treatment and prevention of cancer. tends to be dominated by patient-support type posts. Examples: "Anyone else on drug X?", "Information wanted on colorectal cancer".

sci.med.diseases.hepatitis

Medium volume group on hepatitis diseases. Examples: "Is hepatitis c a sexually transmitted disease?", "Elevated liver enzymes".

sci.med.diseases.lyme

Medium volume group on Lyme disease — patient support, research and information. Examples: "A friend has just been diagnosed", "Lyme and allergies".

sci.med.immunology

High volume group on medical aspects of immunology. Can have a fairly high noise level, partly due to perennial thread on vaccination. (It's all a conspiracy you know. Oh yes.) Examples: "Whole blood lymphocyte assay", "Smallpox vaccination".

sci.med.informatics

Medium volume group on computer applications in medicine. Examples: "Managing medical records", "Computer-aided diagnosis software for Macs wanted"

sci.med.laboratory

Medium volume group on laboratory medicine. Examples: "References needed for arsenic toxicology", "Verifying normal ranges".

sci.med.midwifery

High volume, moderated group on midwifery. Examples: "Information on water births requested", "Epidurals and second stage labour".

sci.med.nursing

High volume group on nursing. A surprising amount of noise. Examples: "International nurses day", "Operating theatre nursing".

sci.med.nutrition

High volume group on medical aspects of nutrition. examples: "Differences between fats", "What is lecithin for?".

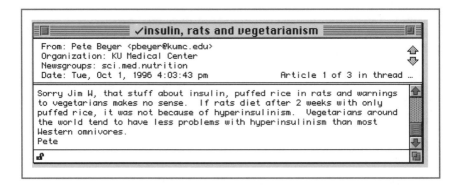

sci.med.occupational

Medium volume group on occupational medicine, including RSI. examples: "Shift-worker mortality rates", "Piano playing and wrist trouble".

sci.med.orthopedic

Low/medium, volume moderated group on orthopaedic surgery and related issues. Examples: "Fracture of internal fixation device", "Functional arm prosthetic".

sci.med.pathology

Medium volume group on disease diagnosis and treatment. Attracts a lot of noise and cross-posts for some reason. Good FAQs. Examples: "Gross specimens", "Ownership of specimens".

sci.med.pharmacy

High volume group on pharmaceutical medicine. Some noise. Examples: "Child-proof caps", "Side effects of drug X".

sci.med.physics

Low volume group on medical aspects of physics. A lot of cross-posting. Examples: "Axial resolution in multislice PET cameras", " Radiation detection".

sci.med.prostate.bph

Low volume group on benign prostatic hypertrophy. Example: "Laser treatment for enlarged prostates".

sci.med.prostate.cancer

Low volume group on prostatic cancer. Example: "Researching new treatments".

sci.med.prostatitis

Medium volume group on prostatitis. Exhaustive FAQ. Examples: "Clinical trial of new treatment", "Rectal spasms".

sci.med.psychobiology

Medium volume group on psychiatry and psychobiology. Examples: "Prozac and lithium for unipolar depression?", "What is a functional psychosis?".

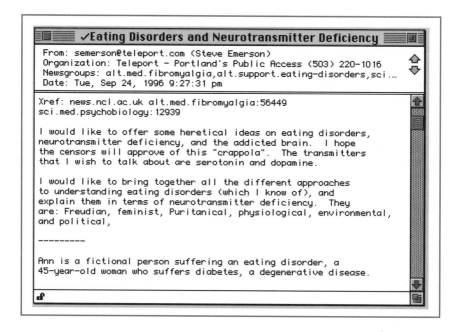

▣▤ ✓Eating Disorders and Neurotransmitter Deficiency ▤▦

```
From: semerson@teleport.com (Steve Emerson)
Organization: Teleport - Portland's Public Access (503) 220-1016
Newsgroups: alt.med.fibromyalgia,alt.support.eating-disorders,sci...
Date: Tue, Sep 24, 1996 9:27:31 pm
```

```
Xref: news.ncl.ac.uk alt.med.fibromyalgia:56449
sci.med.psychobiology:12939

I would like to offer some heretical ideas on eating disorders,
neurotransmitter deficiency, and the addicted brain.  I hope
the censors will approve of this "crappola".  The transmitters
that I wish to talk about are serotonin and dopamine.

I would like to bring together all the different approaches
to understanding eating disorders (which I know of), and
explain them in terms of neurotransmitter deficiency.  They
are: Freudian, feminist, Puritanical, physiological, environmental,
and political,

----------

Ann is a fictional person suffering an eating disorder, a
45-year-old woman who suffers diabetes, a degenerative disease.
```

sci.med.radiology

Medium volume group on radiology. Examples: "How can we tell if tumour cells are dead?", "Investigative radiology".

sci.med.telemedicine

Medium/high volume group on hospital/physician networks. Not a group for on-line diagnosis. High proportion is made up of cross-posts. Examples: "Tele-opthalmology", "Laws on medical software".

sci.med.transcription

High volume group on medical transcription i.e. typing up all those tape recorded mutterings into understandable notes. Could be worse — imagine having to read their handwriting to do it. Examples: "Medical phrase index", " Help requested on word X".

sci.med.vision

High volume group on human vision, visual correction and visual science. Examples: "Progressive bifocals", "Visual field defects in glaucoma".

sci.military.moderated

High volume, moderated group on technical aspects of military hardware. There are a lot of sad people in the world. Take the toys from the boys, we say. Examples: "Camouflage patterns", "Different types of pressure mine".

sci.military.naval

High volume group on past, present and future navies of the world. As above really, except with added rum, sodomy and the lash. Examples: "Dreadnoughts and battle-cruisers", "Exocets vs harpoon missiles".

sci.misc

Medium volume group on general science issues. Suffers from noise and cross-posts. Examples: "What direction does water spiral at the equator?".

sci.nanotech

Medium volume, moderated group on self-reproducing, molecular-scale machines. Discussions are only slightly stymied by the current non-existence of such machines. Examples: "Nanotechnology in everyday life", "Questions about nanotechnology".

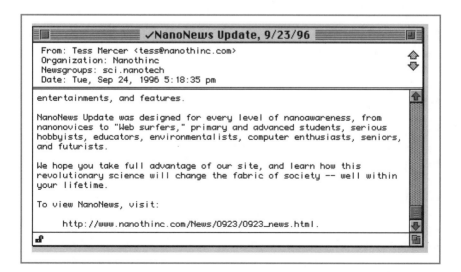

```
✓NanoNews Update, 9/23/96
From: Tess Mercer <tess@nanothinc.com>
Organization: Nanothinc
Newsgroups: sci.nanotech
Date: Tue, Sep 24, 1996 5:18:35 pm

entertainments, and features.

NanoNews Update was designed for every level of nanoawareness, from
nanonovices to "Web surfers," primary and advanced students, serious
hobbyists, educators, environmentalists, computer enthusiasts, seniors,
and futurists.

We hope you take full advantage of our site, and learn how this
revolutionary science will change the fabric of society -- well within
your lifetime.

To view NanoNews, visit:

        http://www.nanothinc.com/News/0923/0923_news.html.
```

sci.nonlinear

Medium volume group on chaotic and other non-linear scientific study. Examples: "Chaos in Turing machines", "Chaotic behaviour in scientific transduction".

sci.op-research

Medium volume group on research, testing and application of operations research. Examples: "Human-aided optimisation", "Constraint solving".

sci.optics

High volume group on optical science. Examples: "Spectroscopy of a candle flame", "Where can I buy X lenses".

sci.optics.fiber

Low volume group on fibre optics components, systems and application.

sci.philosophy.meta

Medium volume group ostensibly about 'meta' philosophies. In practice it is full of noise, trivia and cross-posts.

sci.philosophy.tech

Medium volume group on technical philosophy such as mathematics, science and logic. Has much the same problems as s.p.meta and indeed many of the same posts.

sci.physics

High volume group on all aspects of physics. Tolerable signal to noise ratio — why is this the case with sci.physics and sci.chem but not sci.bio.misc? Examples: "Why is night dark?", "Does light have mass?".

sci.physics.accelerators

Low volume group on particle accelerators and the physics of beams. Examples: "Vector particle geometry", "Medical linear accelerator".

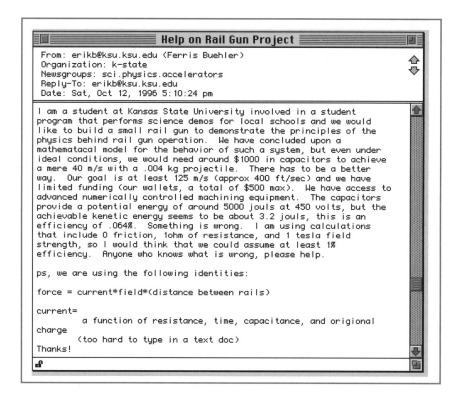

Help on Rail Gun Project

From: erikb@ksu.ksu.edu (Ferris Buehler)
Organization: k-state
Newsgroups: sci.physics.accelerators
Reply-To: erikb@ksu.ksu.edu
Date: Sat, Oct 12, 1996 5:10:24 pm

I am a student at Kansas State University involved in a student
program that performs science demos for local schools and we would
like to build a small rail gun to demonstrate the principles of the
physics behind rail gun operation. We have concluded upon a
mathematical model for the behavior of such a system, but even under
ideal conditions, we would need around $1000 in capacitors to achieve
a mere 40 m/s with a .004 kg projectile. There has to be a better
way. Our goal is at least 125 m/s (approx 400 ft/sec) and we have
limited funding (our wallets, a total of $500 max). We have access to
advanced numerically controlled machining equipment. The capacitors
provide a potential energy of around 5000 jouls at 450 volts, but the
achievable kenetic energy seems to be about 3.2 jouls, this is an
efficiency of .064%. Something is wrong. I am using calculations
that include 0 friction, 1ohm of resistance, and 1 tesla field
strength, so I would think that we could assume at least 1%
efficiency. Anyone who knows what is wrong, please help.

ps, we are using the following identities:

force = current*field*(distance between rails)

current=
          a function of resistance, time, capacitance, and origional
charge
          (too hard to type in a text doc)
Thanks!

sci.physics.computational.fluid-dynamics

Low/medium volume group on computational fluid dynamics. Examples: "Help needed with pressure corrections", "Velocity boundary condition at r = 0?".

sci.physics.cond-matter

Low volume group on condensed matter physics theory and experiment. Mainly taken up with cross-posts.

sci.physics.electromag

Medium volume group on electromagnetic theory and applications. Examples: "Charged particles and magnetic fields", "Conformal transformation".

sci.physics.fusion

Medium volume group on nuclear fusion. Emphasis is on cold fusion, with attendant noise problem. Examples: "Anomalous heat generation", "Cold fusion — does it really happen?".

sci.physics.particle

Medium volume group on particle physics. Some noise. Examples: "Vector particle geometry", "Non-perturbative QFT stats".

sci.physics.plasma

Low volume, moderated group on plasma science and technology. Examples: "Plasma beam attenuation code", "Line broadening data".

sci.physics.research

Low/medium volume, moderated group on physics research. Examples: "Is c really a constant?", "Magnetic field of a magnetised disk".

sci.polymers

Medium volume group on polymer science. Examples: "PTFE heat distortion temperature", "Polyeurethane sponge formulation wanted".

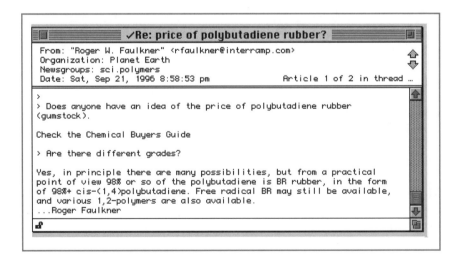

```
┌──────────────────────────────────────────────────────────────┐
│ ▓▓▓▓▓▓▓▓▓▓▓▓ ✓Re: price of polybutadiene rubber? ▓▓▓▓▓▓▓▓▓▓  │
├──────────────────────────────────────────────────────────────┤
│ From: "Roger W. Faulkner" <rfaulkner@interramp.com>       ⬆   │
│ Organization: Planet Earth                                ⬇   │
│ Newsgroups: sci.polymers                                      │
│ Date: Sat, Sep 21, 1996 8:58:53 pm     Article 1 of 2 in thread ...│
├──────────────────────────────────────────────────────────────┤
│ >                                                         ⬆   │
│ > Does anyone have an idea of the price of polybutadiene rubber│
│ (gumstock).                                                   │
│                                                               │
│ Check the Chemical Buyers Guide                               │
│                                                               │
│ > Are there different grades?                                 │
│                                                               │
│ Yes, in principle there are many possibilities, but from a practical│
│ point of view 98% or so of the polybutadiene is BR rubber, in the form│
│ of 98%+ cis-(1,4)polybutadiene. Free radical BR may still be available,│
│ and various 1,2-polymers are also available.                  │
│ ...Roger Faulkner                                         ⬇   │
│                                                               │
│ ⚓                                                             │
└──────────────────────────────────────────────────────────────┘
```

sci.psychology

Obsolete group.

sci.psychology.announce

Low volume, moderated group of announcements relevant to psychology. Consists of announcements on conferences, workshops, job vacancies and other matters of interest to those in the field.

sci.psychology.conciousness

Low volume, moderated group on the nature of conciousness. Examples: "Neurocognitive models of consciousness", "Extrathalamic influences on attention and consciousness".

sci.psychology.digest

Obsolete group.

sci.psychology.journals.psyche

Low volume, moderated group consisting of issues of the electronic journal Psyche, which deals with consciousness.

sci.psychology.journals.psycoloquy

Low volume moderated group consisting of issues of the electronic journal Psycoloquy, which deals with psychology.

sci.psychology.misc

Medium/high volume group on all aspects of psychology. Noisy, as you might expect. Examples: "Things that therapists do wrong", "Illness and behaviour".

sci.psychology.personality

Medium/low volume group on personality and its measurement. Examples: "Personality disorders", "Multiple personalities and being on-line".

sci.psychology.psychotherapy

Medium/high volume group on the practice of psychotherapy. Judging by some of the posts, practitioners should be in work for some time to come. Examples: "Psychotherapy has failed", "Is clinical psychology a science?".

sci.psychology.research

Low volume, moderated group on psychology research. Examples: "Testing of the mentally retarded", "Conducting useful research on the Internet".

sci.psychology.theory

Low/medium volume group on theory of psychology and behaviour. Poor signal to noise ratio — it is almost entirely taken up with noise and cross-posts.

sci.research

Low/medium volume group on all issues related to scientific research. Examples: "Breakthroughs rejected for publication", "Books on scientific writing wanted".

sci.research.careers

Medium volume group on research science as a career. Examples: "Are PhDs worth it?", "What is the job situation like in discipline x".

sci.research.postdoc

Medium volume group on postdoctoral science. A large proportion consists of job openings and people looking for jobs.

sci.skeptic

High volume group dedicated to the debunking of fraudulent and pseudo-science. Great fun. We should all make time to read this one. Examples: "Astrology is nonsense", "How spiritualists prey on the vulnerable".

sci.space.news

Low volume, moderated group on news items related to space science.

sci.space.policy

Medium/high volume group on space policy. Examples: "Has the space program been betrayed?", "Is space exploration worth the money?"

sci.space.science

Medium volume, moderated group on space and planetary science and technology. Examples: "Lifting shuttle external fuel tanks into orbit", "Effect of asteroids on trajectories".

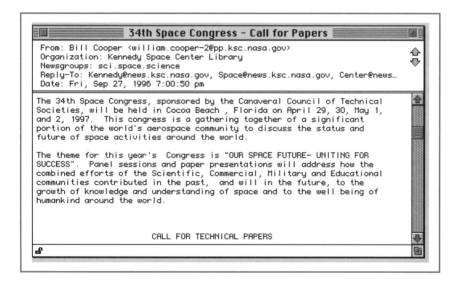

sci.space.shuttle

High volume group about the space shuttle and shuttle transport system (STS) programme. Examples: "Launch postponed", "Deep space exploration and STS".

sci.space.tech

Medium volume, moderated group on technical issues related to space flight. Examples: "Space use of Teflon", "Vacuum welded engines".

sci.stat.consult

Medium volume group on statistical consulting. Many posters take this at face value and post problems here. Examples: " Design and analysis in of computer simulations", "Combining p-values from independent trials".

sci.stat.edu

Medium volume group on statistics education. Examples: "How not to teach statistics", "Student conceptions of probability".

sci.stat.math

Medium volume group on mathematical aspects of statistics. Examples: "Assumptions for regression analysis", "Hypergeometrical distribution".

sci.systems

Low volume group on systems science, theory and support. Examples: "Cycles and systems", "Systems and linking".

sci.techniques.mag-resonance

Low volume group on the use of magnetic resonance imaging and spectroscopy. Examples: "Fibre optical stimulus transmission into MR scanners".

sci.techniques.mass-spec

Low volume, moderated group on mass spectrometry. Example: "Bioanalysis with mass spectrometer".

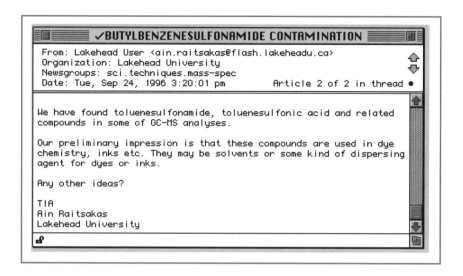

✓BUTYLBENZENESULFONAMIDE CONTAMINATION

From: Lakehead User <ain.raitsakas@flash.lakeheadu.ca>
Organization: Lakehead University
Newsgroups: sci.techniques.mass-spec
Date: Tue, Sep 24, 1996 3:20:01 pm   Article 2 of 2 in thread ●

We have found toluenesulfonamide, toluenesulfonic acid and related compounds in some of GC-MS analyses.

Our preliminary impression is that these compounds are used in dye chemistry, inks etc. They may be solvents or some kind of dispersing agent for dyes or inks.

Any other ideas?

TIA
Ain Raitsakas
Lakehead University

sci.techniques.microscopy

Medium volume group on microscopes and microscopy. Examples: "Buying a microscope", "Scanning photomicroscopy."

sci.techniques.spectroscopy

Medium volume group on spectrum analysis. Examples: "Portable spectrophotometers — any recommendations?", "Protein conformation by spectroscopy".

sci.techniques.testing.misc

Low volume group on testing techniques. Example: "Any ideas on how to test for x?"

sci.techniques.testing.non-destructive

Low/medium volume group on non-destructive tests. Examples: "Phased array ultrasonics", "Measuring the thickness of thin metal layers".

sci.techniques.xtallography

Medium volume group on crystallography. Examples: "Software for crystal shape", "Powder diffraction".

sci.virtual-worlds

Medium volume, moderated group on virtual reality. A good proportion of posts are job ads. Examples: "Virtual worlds with alternative natural laws?", "Best VR helmet/accessories?".

The SOC Hierarchy

soc.culture.scientists

Low volume group dealing with cultural aspects of science. An interesting idea that has never really taken off. Example: "Scientists on the Internet".

soc.history.science

Low/medium volume group dealing with the history of science. Examples: "Gallileo and the church", "Newton — was there really an apple?".

The TALK Hierarchy

This is where the perpetual arguments take place. Talk is a useful place to go to if you want to sharpen your arguments and practice your rhetorical skills. As long as you don't expect to actually change anyone's mind (if anti-scientists were rational they wouldn't hold the views they do, would they?) you might gain something from it. On the other hand, you might feel that life is just too short for this kind of thing. The choice is yours!

talk.environment

High volume group on the environment. Not as bad as you might fear. Examples: "Environmentalism is communism/fascism", "Companies producing environmentally friendly products".

talk.origins

High volume group for the creationism-evolution debate. Seems an absurd idea until you realise that in parts of the USA this is still a live issue. Examples: "Darwin is burning in hell", " How did creationists evolve?".

talk.politics.animals

High volume group full of animal rights rantings and scientific counter-blasts. Examples: "Science is a conspiracy", "Rats are just as important as humans".

The ALT Hierarchy

Here be dragons. The normal rules for newsgroup creation do not apply to alt. This has two effects. Firstly, anyone can create a group on any subject they want, no matter how bizarre. Secondly, alt groups are not as well propagated as the 'Big 8' groups, since they have not passed any test of consumer demand. Alt contains most of the potentially offensive groups on the Internet, much bizarreness and nonsense and some science. In general alt groups tend to have much higher noise than other groups. Alt is a good example of the maxim that fools rush in where angels fear to tread. Consequently, the science-oriented alt groups are of limited use.

alt.agriculture.fruit

Low/medium volume group on fruit production.

alt.agriculture.misc

Medium volume, general agriculture group. Much the same type of posts as sci.agriculture but with more noise and spam.

alt.agriculture.ratite

Empty group on poultry.

alt.architecture

Medium volume group on all aspects of architecture. Examples: "Using CAD", "Use of copper on new buildings".

alt.architecture.alternative

Low/medium volume group on alternative approaches to architecture.

alt.architecture.int-design

Low volume group on interior design.

alt.energy.renewable

Medium volume group on renewable energy resources.

alt.engineering.electrical

Low/medium volume group on electrical engineering. sci.engr.electrical with less on-topic posts and more spam and noise.

alt.engineering.nuclear

Low volume group on nuclear reactor engineering. Few on-topic posts.

alt.engr.explosive

Medium volume group on making bombs. Really.

alt.folklore.science

Medium volume group on science folklore. Anecdotes, stories, urban myths and general science questions.

alt.human-brain

Low volume group with little or no posts relevant to the human brain.

alt.med.allergy

Low/medium volume group on allergies.

alt.med.cfs

High volume group on chronic fatigue syndrome. Regularly posted charter and FAQ keeps the group well-focussed and on-topic. Gated to the CFS-l mailing list.

alt.med.equipment

Low volume group on medical equipment — mainly wanted or for sale.

alt.med.fibromyalgia

High volume group on fibromyalgia research, treatment and support. Mainly patient support.

alt.med.phys-assts

Low volume group for physical assistants

alt.med.urum-outcomes

We can't work out what this is supposed to be about. There weren't any posts in it anyway.

alt.med.veterinary

Medium volume group on animal health and medicine.

alt.med.vision.improve

Empty group on improving vision.

alt.medical.sales.jobs.offered

Low volume group of job ads for sales reps.

alt.psychology

Medium volume psychology group. Similar to sci.psychology, but with more noise.

alt.psychology.alderian

Low volume group on Alderian psychology.

alt.psychology.help

Medium volume general psychology group.

alt.psychology.jung

Low/medium volume archetypal psychology group.

alt.psychology.mistake-theory

Low volume group on mistake theory. All cross-posts.

alt.psychology.nlp

Medium volume group on neuro-linguistic programming.

alt.psychology.personality

medium volume group on the psychological aspects of personality.

alt.psychology.transpersonal

Low volume group on transpersonal psychology.

alt.sci.astro

All of these groups have a low volume and contain few if any astronomy posts. Superceded by the sci.astro groups.

alt.sci.image-facility

Low volume group on image facility.

alt.sci.joe-bay

Don't know who he is/was, or why he has his own newsgroup, but this is a low volume group full of spam anyway.

alt.sci.physics.acoustics

Low/medium volume group on acoustics. Appropriately enough it has a good signal to noise ratio. Examples: "Directionality of musical instruments", "SPL meters".

alt.sci.physics.new-theories

Medium volume group on new theories in physics i.e. high wackiness content.

alt.sci.physics.plutonium

Medium volume group which gives a home to Archimedes Plutonium and his many theories.

alt.sci.physics.spam

Low volume group which hasn't even managed to attract much spam.

alt.sci.planetary

Low/medium volume group on planetary science. Mainly cross-posts from the sci.astro groups.

alt.sci.sociology

Low/medium group on sociology. As if sci.economics wasn't bad enough.

alt.sci.tech.indonesia

Low volume group on Indonesian science and technology.

alt.sci.time-travel

Low volume group on time-travel.

alt.technology.misc

Low volume group on technology.

alt.technology.mkt-failure

Low volume group about technology market failures.

alt.technology.obsolete

Low volume group on obsolete technology.

alt.technology.smartcards

Low volume group on smartcards

The BIONET Hierarchy

A Usenet oasis of sanity for biologists, Bionet is a hierarchy of newsgroups on all aspects of biological research, intended for scientists only. It has its own, simplified, rules for newsgroup creation as described earlier. Better still, every single bionet newsgroup is also available as an email mailing list, for those who do not have access to newsgroups. The way to subscribe by email depends on where in the world you live. This is because Bionet is run from 2 nodes, one US based (for the Americas and the Pacific Rim countries) and one UK based (for Europe, Africa and Central Asia)

You can subscribe to either node wherever you live. However it will be faster, and help to reduce traffic on the already overloaded transatlantic links, if you use the node in your own hemisphere. Being a good netizen, you will no doubt do exactly that.

Confusingly, the method for subscribing is different for each node. For the UK node, send the message:

sub bionet-news.<complete name of newsgroup>

to the address:

mxt@dl.ac.uk

So to subscribe to bionet.diagnostics, you would send the message:

sub bionet-news.bionet.diagnostics

For the US node, you need to send the message:

subscribe <name of mailing list>

to the address:

biosci-server@net.bio.net

So, to take the example of bionet.diagnostics again, you would send the message:

subscribe diagnost.

Now hold on, you say, how am I supposed to know that 'diagnost' was the name to use for subscribing to bionet.diagnostics? Well, we could see that that could cause some confusion and so we have included these mailing list names in the listing of bionet newsgroups below. Remember you only need to use them if you are covered by the US node.

We haven't bothered adding any comments about 'noise' — because bionet newsgroups don't have any significant noise. The tiny (by Usenet standards) amount of residual spam has led to an increase in the number of moderated groups. The volume levels need to be recalibrated (remember what we said about less being more?). Low volume means less than 50 posts per month, medium is 51–150 and high volume is over 150. There is one very high group, bionet.molbio.methds_and_reagnts, which has well over 1000 posts per month — probably one to avoid subscribing to by email if you can avoid it.

bionet.agroforestry

High volume newsgroup on agroforestry research. Mailing list name: ag-forst.

bionet.announce

Medium volume, moderated newsgroup consisting of announcements of interest to biologists. Mailing list name: bionews.

bionet.audiology

Medium volume newsgroup about audiology and hearing science. Mailing list name: audiolog.

bionet.biology.cardiovascular

Low volume cardiovascular research discussions. Mailing list name: cardiors.

bionet.biology.computational

Low volume newsgroup on mathematical and computer applications in biology. Mailing list name: comp-bio.

bionet.biology.deepsea

Low volume newsgroup on research in deep-sea marine biology, oceanography and geology. Mailing list name: deepsea.

bionet.biology.grasses

Low volume newsgroup on grasses, especially cereal, forage and turf species. Mailing list name: grasses.

bionet.biology.n2-fixation

Low volume newsgroup on biological nitrogen fixation. Mailing list name: n2fix.

bionet.biology.symbiosis

Low volume newsgroup on symbiosis research. Mailing list name: symbios.

bionet.biology.tropical

Low volume newsgroup on research in tropical biology. Mailing list name: trop-bio.

bionet.biology.vectors

Low volume, moderated newsgroup on research and control of arthropods which spread disease. Mailing list name: vect-bio.

bionet.biophysics

Medium volume group on biophysics research. Mailing list name: biophys.

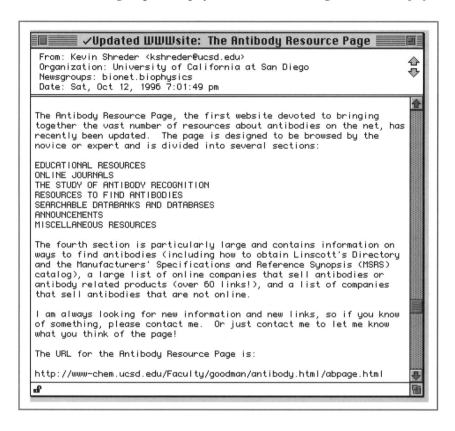

bionet.celegans

Low volume research discussions on *Caenorhabditis elegans* and related nematodes. Mailing list name: celegans.

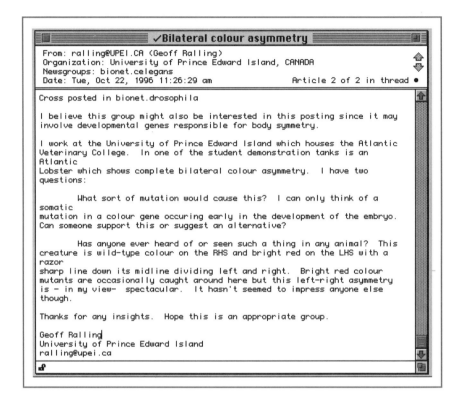

bionet.cellbiol

High volume newsgroup on cell biology, including cancer research at the cellular level. Mailing list name: cellbiol.

bionet.cellbiol.cytonet

Low volume newsgroup on research on the cytoskeleton, plasma membrane, and cell wall. Mailing list name: cytonet.

bionet.cellbiol.insulin

Low volume, moderated group about the biology and chemistry of insulin and related hormones and receptors. Mailing list name: insulin.

bionet.chalamydomonas

Low volume newsgroup about biology of the green alga *Chlamydomonas* and related genera. Mailing list name: chlamy.

bionet.diagnostics

Medium volume discussion of problems and techniques in all fields of diagnostics. Mailing list name: diagnost. This is Kevin O'Donnell's favourite newsgroup — not that he's at all biased just because he moderates it.

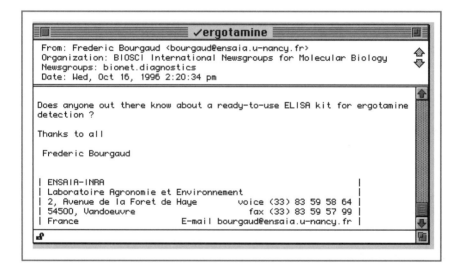

bionet.diagnostics.prenatal

Low volume newsgroup concerning research in prenatal diagnostics. Mailing list name: prenatal.

bionet.drosophila

Medium volume newsgroup on biological research in Drosophila. Mailing list name: dros.

bionet.ecology.physiology

Low volume, moderated, discussions about research and education in physiological ecology. Mailing list name: ecophys.

bionet.emf-bio

Low volume, moderated newsgroup on research on electromagnetic field interactions with biological systems. Mailing list name: emf-bio.

bionet.general

Very high (5–600) volume newsgroup consisting of miscellaneous biology discussions. Mailing list name: bioforum.

bionet.genome.arabidopsis

Medium volume newsgroup on the arabidopsis genome project. Mailing list name: arab-gen.

bionet.genome.autosequencing

Low volume, moderated newsgroup about automated DNA sequencing. Mailing list name: autoseq.

bionet.genome.chromosomes

Low volume newsgroup on mapping and sequencing of eukaryote chromosomes. Mailing list name: biochrom.

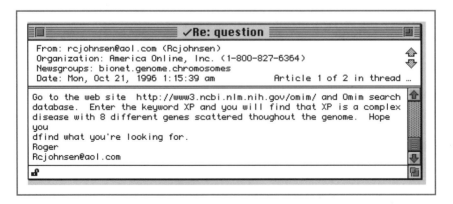

bionet.glycosci

Low volume newsgroup concerning carbohydrate and glycoconjugate molecules. Mailing list name: glycosci.

bionet.immunology

High volume newsgroup on immunology research. Mailing list name: immuno.

bionet.info-theory

A medium volume newsgroup on application of information theory to biology. Mailing list name: bio-info.

bionet.jobs.offered

High volume, moderated newsgroup consisting of job opportunities in biology. Mailing list name: biojobs.

bionet.jobs.wanted

Very high volume newsgroup (500) which provides a forum for posting resumes/CVs by individuals seeking employment in the biological sciences or in support of the biological sciences. Mailing list name: wantjob.

bionet.journals.contents

Low volume newsgroup consisting of tables of contents of biological journals — often before the actual publication date. Mailing list name: bio-jrnl.

bionet.journals.letters.biotechniques

Low volume newsgroup on discussions of articles in the journal Biotechniques. Mailing list name: btechniq.

bionet.journals.letters.tibs

Low volume newsgroup consisting of "Letters to the Editor" of Trends in Biochemical Sciences". Mailing list name: tibs.

bionet.journals.notes

Low volume newsgroup on practical advice on dealing with professional journals. Mailing list name: jrnlnote.

bionet.metabolic-reg

Low volume newsgroup on kinetics and thermodynamics at the cellular level. Mailing list name: btk-mca.

bionet.microbiology

High volume discussions about microbiology research. Mailing list name: microbio.

bionet.molbio.ageing

Medium volume newsgroup on ageing research. Mailing list name: ageing.

bionet.molbio.bio-matrix

Low volume discussions about applications of computers to biological databases. Mailing list name: biomatrix.

bionet.molbio.embldatabank

Low volume newsgroup consisting of messages to and from the EMBL staff. Mailing list name: embl-db.

bionet.molbio.evolution

Medium volume discussions about research in molecular evolution. Mailing list name: mol-evol.

bionet.molbio.gdb

Low volume newsgroup consisting of messages to and from the Genome Data Bank staff. Mailing list name: gdb.

bionet.molbio.genbank

Low volume newsgroup consisting of messages to and from GenBank database staff. Mailing list name: genbankb.

bionet.molbio.gene-linkage

Low volume newsgroup on genetic linkage analysis. Mailing list name: gen-link.

bionet.molbio.genome-programme

Low volume, moderated, NIH-sponsored newsgroup on human genome issues. Mailing list name: gnome-pr.

bionet.molbio.hiv

Medium volume newsgroup on molecular biology of HIV. Mailing list name: hiv-biol.

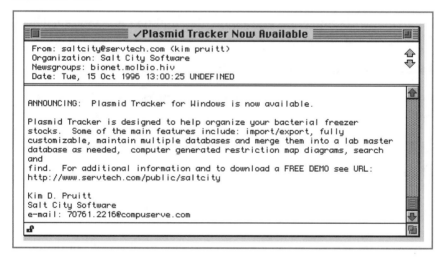

bionet.molbio.methds-reagents

Very high volume (1400+) newsgroup dealing with requests for information on methods and lab reagents. Mailing list name: methods.

bionet.molbio.mollusc

Low volume, moderated newsgroup on research on mollusc DNA. Mailing list name: molluscs.

bionet.molbio.proteins

High volume discussions about research on proteins and messages for the PIR and SWISS-PROT databank staffs. Mailing list name: proteins.

bionet.molbio.proteins.fluorescent

Medium volume newsgroup on fluorescent proteins and bioluminescence. Mailing list name: fluorpro.

bionet.molbio.proteins.7tms_r

Low volume newsgroup on signal transducing receptors which interact with G-proteins. Mailing list name: 7tms_r.

bionet.molbio.rapd

Low volume discussions about randomly amplified polymorphic DNA. Mailing list name: rapd.

bionet.molbio.recombination

Low volume newsgroup on research on recombination of DNA or RNA. Mailing list name: recom.

bionet.molbio.yeast

High volume newsgroup on molecular biology and genetics of yeast. Mailing list name: yeast.

bionet.molecules.peptides

Low volume newsgroup on chemical and biological aspects of peptides. Mailing list name: peptides.

bionet.molec-model

Low volume discussions on molecular modelling. Mailing list name: molmodel.

bionet.molecules.repertoires

Low volume newsgroup on generation and use of libraries of molecules. Mailing list name: molreps.

bionet.mycobiology

High volume newsgroup on research on filamentous fungi. Tends to attract more laypeople than other bionet groups. Mailing list name: mycology.

bionet.neuroscience

Very high volume (500) newsgroup on neuroscience research. Mailing list name: neur-sci.

bionet.neuroscience.amyloid

Low volume newsgroup on research into Alzheimer's disease and related disorders including prion diseases. Mailing list name: amyloid.

bionet.organisms.pseudomonads

Low volume discussions on research on the genus *Pseudomonas*. Mailing list name: pseudomo.

bionet.organisms.schistosoma

Low volume newsgroup on Schistosoma research. Mailing list name: schisto.

bionet.organism.urodeles

Low volume, moderated newsgroup on research in urodele amphibian biology. Mailing list name: urodeles.

bionet.organisms.zebrafish

Low volume discussions on research using the model organism Zebrafish (*Danio rerio*). Mailing list name: zbrafish.

bionet.parasitology

Low volume newsgroup on parasitology research. Mailing list name: parasite.

bionet.photosynthesis

Low volume, moderated newsgroup on photosynthesis research. Mailing list name: photosyn.

bionet.population-bio

Low volume newsgroup on research in population biology. Mailing list name: pop-bio.

bionet.plants

High volume newsgroup on plant biology research. Mailing list name: plantbio.

bionet.plants.education

Medium volume discussions on education issues in plant biology. Mailing list name: plant-ed.

bionet.plants.signaltransduc

Newsgroup on plant signal transduction. Mailing list name: plsignal.

bionet.prof-society.afcr

Low volume, moderated newsgroup consisting of American Federation for Clinical Research announcements. Mailing list name: afcr.

bionet.prof-society.aibs

Moderated newsgroup consisting of matters relevant to the American Institute of Biological Sciences. Mailing list name: aibs.

bionet.prof-society.ascb

Low volume, moderated newsgroup consisting of American Society for Cell Biology announcements. Mailing list name: ascb.

bionet.prof-society.biophysics

Low volume, moderated newsgroup consisting of announcements / information from the Biophysical Society. Mailing list name: bphyssoc.

bionet.prof-society.cfbs

Medium volume, moderated newsgroup on matters relevant to the Canadian Federation of Biological Societies. Mailing list name: biocan.

bionet.prof-society.csm

Low volume, moderated newsgroup on matters relevant to the Canadian Society of Microbiologists. Mailing list name: csm.

bionet.prof-society.faseb

Low volume, moderated newsgroup for announcements from the Federation of American Societies for Experimental Biology. Mailing list name: faseb.

bionet.prof-society.navbo

Low volume, moderated newsgroup for the North American Vascular Biology Association. Mailing list name: navbo.

bionet.protista

Low volume newsgroup on research on ciliates and other protists. Mailing list name: protista.

bionet.sci-resources

Low volume, moderated newsgroup consisting of information from or about science funding agencies. Mailing list name: sci-res.

bionet.software

High volume newsgroup on information on software for the biological sciences. Mailing list name: bio-soft.

bionet.software.acedb

Low volume newsgroup dedicated to discussions by users and developers of the ACEDB software used in genome databases. Mailing list name: acedb.

bionet.software.gcg

Low volume newsgroup on GCG sequence analysis software. Mailing list name: info-gcg.

bionet.software.srs

Low volume discussions about the Sequence Retrieval System (SRS) software. Mailing list name: bio-srs.

bionet.software.staden

Low volume newsgroup on Staden sequence analysis software. Mailing list name: staden.

bionet.software.x-plor

Medium volume newsgroup on x-plor software for 3D molecular structure determination. Mailing list name: x-plor.

bionet.software.www

Low volume newsgroup on announcements about new biology resources on the WWW. Mailing list name: bio-www.

bionet.structural-nmr

Medium volume newsgroup on the use of NMR for structure determination. Mailing list name: str-nmr.

bionet.toxicology

Medium volume newsgroup on toxicology research. Mailing list name: toxicol.

bionet.users.address

Low volume newsgroup providing support on using networks and finding email addresses. Mailing list name: bio-naut.

bionet.virology

High volume newsgroup on virology research. Mailing list name: virology.

bionet.women-in-bio

High volume newsgroup on issues concerning women biologists. Mailing list name: womenbio.

bionet.xtallography

Medium volume newsgroup for discussion about crystallography of macromolecules and messages for the PDB staff. Mailing list name: xtal-log.

The K12 Hierarchy

Newsgroups for school teachers.

k12.ed.math

Medium volume group on maths education. Examples: "Use of the abacus", "How to calculate the area of a parrallelogram?".

k12.ed.science

Medium volume group on science education. Examples: "New science education web page", "Why does this experiment never work?".

Help Others

Have we missed your favourite newsgroup — or worse, are our views on it completely wrong? Write and tell us and we'll put the information on the "erratum" web site.
<<http://www.compulink.co.uk/~embra/ifs.html>>

Real Time Chat
MUDs, MOOs, IRCs, Electronic Conferencing

Nease Androj has just discovered multi-user discussion groups. As a single postdoc, Nease has extra time late in the evenings after the lab technicians have gone home, so visiting BioMOO just to listen in on the conversations happening there has become a regular educational enterprise. Nease has always been shy, and yet very alert, so understanding the thread of a conversation isn't too much of a problem. While the newsgroups are a slow-moving kind of forum, the MUD is altogether quicker, with immediate feedback when one or the other party gets bored. Also, these real-time discussions bring the participants that much closer together, although they're a bit frustrating if you can't type very fast. Nease prefers the cloak of anonymity in these discussions, because questions can be asked without making a fool of one's 'real' self, risking only one's transient virtual identity. Of course, one never knows the provenance of advice in these sessions either, but the trail initiated there could prove fruitful on continued search. And Nease is growing in confidence that useful questions can be asked, in public forums — just yesterday a thought developing out of a MUD occurred during the lab meeting that, verbalised aloud, elicited frank curiosity and further interest from the Prof!

Explanation of Abbreviations:

MUDs: Multiple User Dimensions / Devices
MOOs: Multiple user Object Oriented devices
IRCs: Internet Relay Chat

WHAT'S A CHAT GROUP?

A chat group on the Internet is where different PCs all around the world are hooked up via a central server (we've used this term quite a bit throughout the book), so that when one of them keys in some text on their keyboard, and strikes the send key, each of the other members of the group, sitting actively at their computers, receives the same text. In this way, people converse with each other, by keypunching. And this is progress?

We can't avoid, in our general discussion of the tools of the Internet, a brief nod to some rather more experimental approaches for communication on the net. Some folks call these approaches 'virtual reality', but that's too highfalutin' for us. They're really just hopped-up Internet Chat groups, but 'virtual reality' sounds better. It makes it seem like you are at the cutting edge of electronic communication rather than the sort of person whose idea of a good time is a game of Dungeons and Dragons or who thought that CB radio was really good idea. [*'Ere, I thought CB was a good idea, though I never did buy one* — **LW**] When the popular media (say the tabloids) go hyper about virtual reality on the Internet, it's these MUDs / MOOs / IRCs that they're going gaga over. Here's an example of one such space, as found in '**itropolis**' (as facilitated with the '**ichat**' application) complete with a character present to talk to:

```
Welcome to Chat Alley.
Obvious exits: World_View, Moes_Bar and Rocket_Cafe
You see here: Jonbredo
```

Now it's up to you to 'speak' to Jonbredo, or otherwise communicate with this person. In this case, the landscape of the 'Chat alley' remains the same, while the text window below changes, but in other 'virtual reality' spaces, you do see different pictures depending on which 'room' you're in. In most cases, however, you communicate by hitting those old keys, and you 'send' your message by the return key. There are lots of Internet Chat spaces and servers, and it's a great old world, we guess, of tap-tap-tap talk under assumed personae. Oh, by the way (btw), we're going to let you find the '**itropolis**' space yourself.

One day, virtual reality conferences will allow you to present impressive multimedia presentations to scientists from all over the globe (apart from those who are in their beds — remember, it's always the middle of the night *somewhere*) in real time — and allow them to interact in equally rapid and impressive ways. However, we're not quite there yet, though a few brave souls have attempted to claim this technology for science.

It's true, we're familiar with at least a couple of virtual electronic conferences in the scientific domain, which have left their virtual attendees somewhat virtually bemused. The power of these MUDs, or MOOs, is that you can converse on the keyboard with colleagues, in real time. So that when you type things out, and hit the send button, your contribution to the conversation (this is all filtered through appropriate software, of which an example appears below) is instantly added to the flow that appears on your screen. Sounds good in theory, but in practice, contributors who are otherwise articulate orally usually find they become tongue-tied at the keyboard. Probably because they're still typing with two fingers. But what that means is that discussions become rather laboured.

Here we include a small conversation, or chat, between me, myself and I, as realised using the seamlessly integrated 'Private Chat' application in the Open University's First Class network. This sequence should demonstrate the basic principle of a chat between like-minded colleagues; in the first panel, I have initiated the application, and the window opens up.

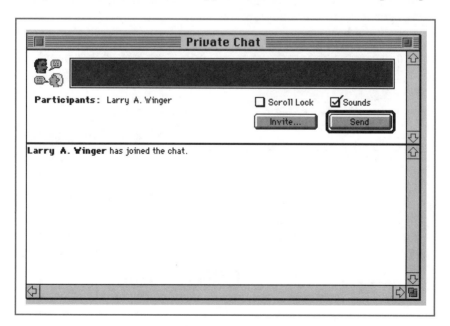

Now I need to select somebody with whom to chat. Let's see, who else is online? Oh, well, I am, why don't I have a chat with somebody who really understands me?

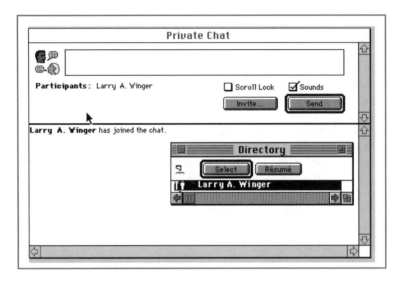

Okay, let's invite Larry A. Winger into the chat, then. Being online, Larry A. Winger receives the following message:

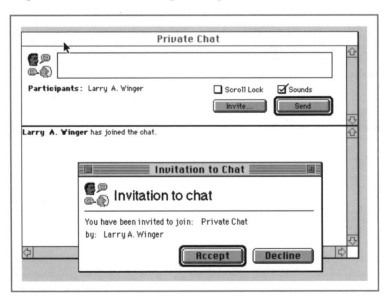

When I accept, I join the chat. Seamless! I key in a few strokes.

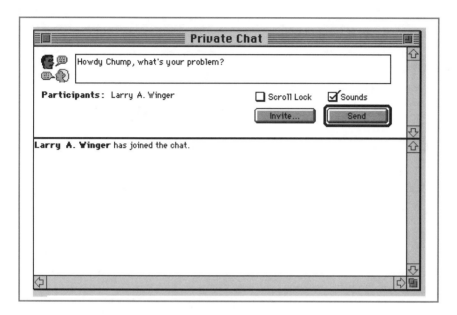

When I hit the 'send' button, the message appears in the active 'chat' space seen by both participants.

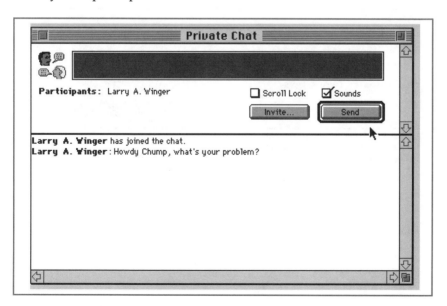

And I can then respond to my query:

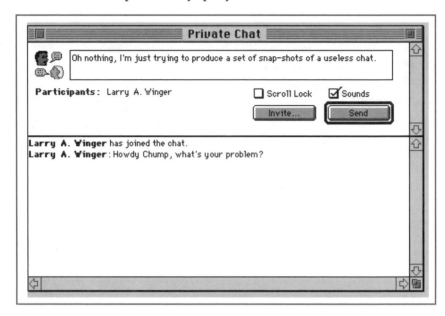

We hope you get the picture:

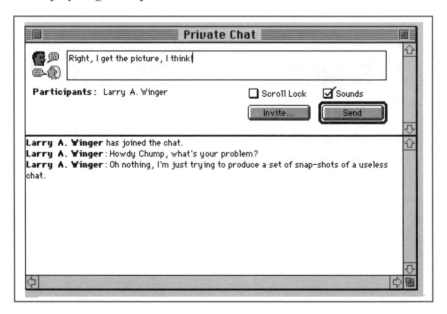

This all seems very straightforward, really, so what's the problem?

Well, the problem is that, in practice, you see, both contributors are laboriously typing in their important contribution on their keyboards, but then one just gets in (ie, hits the return or enter key) first. Hmmmph. And all that time the other had put in making the same point. Now, do they really want to finish and send in the same thing? Will they then be seen to be making a copy-cat ass of themselves? So because of this keyboard gap, contributions tend to be quick, rapid-fire statements which leave you gasping for nuance and detail — see the example further along in the chapter. IRC is heavily used, we're well aware, recreationally, for long-distance chats between lonely nerdies, or self-preoccupied trans-gender iconoclasts, but they've got well developed keyboard skills after all, owing to long lonely hours gazing deeply into their screens. In the final analysis, busy scientists probably just don't have enough time to spend on this sort of remote exercise in frustration — dancing fingers just don't keep up with rapid brains anxious to communicate, at least at the speed of sound! And the embarrasment factor is always quite high — is your spelling skill commensurate with your verbocity?

Of course, by the time you read this chapter, the technology may have advanced to a much more simple and user-friendly stage, and we might be communicating by word of mouth into our very ears, as seen by our very eyes! At the moment though, we reckon that this sort of real-time keyboard communication is for serious net-heads who want to do it this way, rather than a resource that can be exploited by the non-techie user, which is what the rest of this book is about.

Perhaps not satisfactory then, and perhaps we're all really just waiting for full-scale electronic video-conferencing, which is just on the horizon as we write, though the bandwidth requirements for large-scale use will be quite formidable. But certainly between medical centres we're seeing surgical operations broadcast to observers far remote from the actual theatre, as well as the potential for operations actually manipulated re-motely. And we admit it, we've manipulated the archaeological robot at USC, though we haven't watered the flowers there yet. (*I've driven the trains at ULM, LW*). If these comments are baffling — as they probably are — all is explained in the 'Tea-Break' chapter.

So anyway, let's wait for that wonder of the millenium, shall we, and concentrate on information exchange as we want it, and when we want it, which, after all, is why email is rapidly enhancing telephone commu-nication as an increasingly more polite way of communicating. We want the Internet to work for us, after all; we do not want to sit, maddened in our chairs, working for the Internet, do we?

Sure, the future is here, and it's coming too, (and we'll have our cake and eat it) but for now, let's concentrate on the present reality and utility of these wonderful tools, as we consider where and how best to use them, without getting lost in their authors' hype, okay? For example, you can find a Windows-based chat facility which will translate your intelligent monkey-like key tapping into reasonably coherent audible sounds at the other end. But, is this an advantage in chat over direct oral/aural communication? When you can speak into your computer, and have that speech codified, broken down into an acceptable band-width size and relayed into a real-time chat with electronic translation back into aural mode, as is done in old-fashioned analogue conference calls, then maybe Internet chat will be getting somewhere. By that time, of course, direct electronic video conferencing will be with us as well. For most of us mortals, however, stuck with only a screen and a keyboard, the present reality is just not good enough to warrant much effort.

* * *

Okay, if you insist, all right then, here are a couple of examples of scientific electronic conferences or IRCs that are going on as we write. When these conferences in real time are combined with World Wide Web access to a poster presentation, as we discuss in chapters below, they can add a certain piquancy, and they do show us where the technology is leading! Or when a bunch of like-minded individual scientists schedule to meet online to discuss protein structure, as we extract below, one can appreciate something of a useful point.

BIOMOO

Excerpt from the BioMOO FAQ:

"BioMOO is a virtual meeting place for biologists, connected to the Globewide Network Academy. The main physical part of the BioMOO is at the BioInformatics Unit of the Weizmann Institute of Science, Israel. BioMOO is a professional community of Biology researchers. It is a place to come meet colleagues in Biology studies and related fields and brainstorm, to hold colloquia and conferences, to explore the serious side of this new medium."

Further along in this chapter we've got an example of the idea of using the WWW browser to visualise the 'rooms' you enter in a virtual reality like BioMOO, but you may appreciate knowing now that BioMOO has an interface on the WWW where many questions may be answered. To reach it, point your web browser to:

http://bioinfo.weizmann.ac.il/BioMOO

For access to the real virtual chat/conference area, we can do no better than to quote Iddo Friedberg's plangent instructions to new users:

> "The usual client/server communication architecture that serves us on the Internet applies here as well. YOU, sitting there on your Mac/PC/ whatever are a client. The computer running the MOO software through which you wish to talk to other people is the server. What you need right now is a bit of software, running on the client (that is YOUR machine) that will connect you to the server. This bit of software, amazingly, is called a MUD client. Actually, you can also use your telnet software to log into a MOO. Remember, that for most MOOs you should use a different IP port number for telneting. An IP port number is, in effect, the "door" via which you "walk" into the computer hosting the MOO you would like to get through to. Since that computer may be used for purposes other than MOOing, those in charge of that computer have decided to allot a "guest entrance" for MOOers. On the BioMOO, the port number is 8888. The address for telneting is:

bioinfo.weizmann.ac.il

To get into the BioMOO simply type:

> * On a Mac: use NCSA telnet (or any other telnet software you might have). Type in the session name as bioinfo.weizmann.ac.il Port number will be 8888.
> * On a PC: use the telnet software you're used to. Make sure that you typed in 8888 in the IP port field."

It's true that telnet software by itself feels clunky in IRC, with lots of repeated messages as you type something in, and then see it repeated when you've sent it along to the server. This is why we've carefully demonstrated the special 'client program' called 'Private Chat' above. From Iddo Friedberg again, and you might also like to consider our more exhaustive consideration of file transfer in the FTP Chapter:

> "... there are special client programs you can use for MOOing. These have an advantage over simple telneting in that they enable you to reduce screen clutter by separating your outgoing messages from other people's incoming messages, define keyboard shortcuts & macros for often-repeated operations, save sessions locally, and lots of other stuff.

> * For PC/MS-Windows users:
> http://www.ccs.org/winsock/mud.html
> contains a list of many clients you can download and use.

> * For Macintosh users: MUDDweller.
> ftp://mac.archive.umich.edu/mac/util/comm
> ftp://ftp.tcp.com/pub/mud/Clients/

We thought it might be useful to include below a wee transcript of a virtual reality meeting in BioMOO among scientists (students, of course, but aren't we all?), who are trying to figure out how to use the various world databases on proteins. Remember, everybody in this meeting is seeing these text queries and responses in simultaneous mode, as soon as the sender hits the return key.

Transcript of meeting in **BioMOO**, PPS Base 14th Mar '96 17:00 GMT on **Protein   Assignment 1**

Thanks to Peter Murray-Rust for holding this well-attended meeting, and for dealing with so many questions.

Transcript

* Discussion on assignment questions 1 and 2 : related proteins and PDB codes
* .... Q3: small molecules in the structure
* .... Q4: large non-protein molecules
* .... Q5: disulphide bonds
* .... Q6: ambiguity of composition
* .... Q7: number of (macro)molecules in the structure
* .... Q8: discrepancies between SEQRES and ATOM records
* .... Q9: oligomeric structures
* Q10–18

PeterMR turns the recorder on.

Giovanni finds his way in.

Ahotz says, "hi all"

Franco says, "hello everybody"

Paolo finds its way in.

PeterMR says, "the recorder is now on"

Paolo waves

Salim says, "Hello to the new commers"

JohnW asks, "does everyone have the appropriate URL on a WWW browser?"

PeterMR says, "we are going to discuss protein assignment 1."

PeterMR says, "has everyone got the assignment"

Giovanni says, "I got the assignment"

Jzt finds its way in.

Luis says, "I got it"

PeterMR says, "any problems with Q1? — similar proteins?"

Salim says, "I have the correct browser on"

Franco says, "I got it"

KarlS says, "I got it too"

Auroram says, "yes I have started working with the assignment"

Paolo says, "Me too"

JohnW says, "thats .../PPS2/assignments/proteins1.html"

PeterMR says, "does everyone have a protein of the sort 1ABC?"

Luis says, "How do you know if a protein is similar, just from the pdb code?"

PeterMR says, "1abc and 2abc are usually related"

Franco says, "what about 1ccr and 1ccx?"

KarlS says, "I found other members of 'my' protein family, but the pdB codes are not similar at all"

PeterMR says, "1ccr and 1ccx are not necessarily related"

PeterMR says, "the prob is the PDB started with only 4-letters, so really the codes don't mean very much."

Silk materializes out of thin air.

PeterMR says, "the early ones, e.g. 1ins for insulin were OK"

PeterMR ghashas a bad deelet keey

Silk aka. Matthew Ellis waves hello

PeterMR says, "there are about 100 structures of mutant lysozymes."

Marek finds its way in.

PeterMR says, "so it is impossible to use similar codes."

TRex says, "Hi — back again"

The housekeeper arrives to cart Silke off to bed.

PeterMR says, "I shall go to another terminal IRL — 1 minute — chat among yourselves . . ."

Ahotz says, "me too"

Auroram [to so]: we have to search for homologies in the sequence to find the possible family members?

Jzt has disconnected.

Tday finds its way in.

Paulyta says, "At some sites you can key-word browse PDB files"

JohnW [to auroram]: one way of finding other family members is to search a database of structures, e.g. there is a search interface at PDB

Tday says, "hello"

PeterMR is back again

We think it's not for the faint-hearted, but it's undeniable that some information is being transferred here. Our question is, how much time do you have to use re-learning how to talk through your fingers?

* * *

In a different sort of approach, electronic conferences are also a blend of IRC technology and WWW pages, but they seek to capture the flavour of a real conference. I [LW] had an interesting time exploring the conference centre from the 1st Electronic Glycoscience Conference, but you really had to register for the second one to experience its pleasures in glowing colour.

SECOND ELECTRONIC GLYCOSCIENCE CONFERENCE

It's a sad, if inevitable fact that there are quite a few electronic conferences on themes of particular interest to computing science professionals, but not so many yet in other areas. The Glycoscience conferences are now well established, however, and in addition the proceedings are published in hard copy journals. As noted on their WWW page (address below) information on this conference is also spread about via bionet newsgroup and mailing list, from whence came this announcement:

"The Second Electronic Glycoscience Conference (EGC-2) will be held on the Internet (the Net) and World Wide Web (the Web) from Sept 9 — 20, 1996 and will follow the same pattern as the first such conference, EGC-1, held in 1995. The conference is sponsored by Chapman & Hall and Oxford Glycosystems.

EGC-2 will be a fully international event open to all members of our scientific community and will cover a broad range of disciplines related to carbohydrate and glycoconjugate molecules including chemical, physical, biological and medical areas using theoretical, experimental and computational approaches.

Further details will be given in the authors' guide accessible via the URL:

http://bellatrix.pcl.ox.ac.uk/egc2/

During the conference discussions will take place via the Internet in real-time using a virtual conference centre based on a MOO (multiple-user domain, object oriented) and via Internet-accessible electronic mailing lists. Trial sessions for those not familiar with MOO will be held before the conference. During the conference, a timetable for MOO discussion sessions of each section will be posted. Since these realtime discussions are an integral part of the conference, authors will be expected to attend one for their subject; the right is reserved not to referee submissions by authors who do not attend one of these sessions.

The Conference will feature a Virtual Trade Center where commercial vendors, consultants, manufacturers, and contractors will be able to display their goods and services in return for exhibition fees to support conference activities. Any potential advertisers should contact the conference organisers.

The Third Electronic Glycosciences Conference will be held, from Monday the 6th October to Friday the 17th October, 1997: check out the URL: http://www.vei.co.uk/egc3/

Somehow, the virtual reality electronic conference just doesn't have the 'je ne sais quoi' of a real conference in an exotic locale, though I (LW) can hardly talk, as I haven't been to a real conference in over 5 years. [Blue Peter have been informed!] So for someone like me, without any travel money available, but with serious keyboard skills, and direct hookup to the Internet via university connection, this sort of conference could be a real god-send, in that abstracts could get published without needing to pay vast hotel bills. So where's the next electronic conference in my own (diagnostics/immunoassay) field? It could pay to keep my eyes open in appropriate newsgroups!

* * *

And finally, we don't want to neglect science teachers, the backbone of our academic community. It's certainly true that you couldn't say that primary and secondary teachers are well advantaged when it comes to conferencing facilities and funding, now could you? Could this sort of conference shown below be useful to teachers? You'd have to check it out to see, and be prepared to have spent a certain amount of time online. How healthy are your telephone resources? Or can you stay up all UK-night meeting colleagues in real American day-time?

Diversity University MOO invites everyone with telnet capabilities to join us as we join the hosts of the FIRST EVER NASA Mars "Virtual" Teachers' Conference! This is a one-day online expansion of a three-day physical conference hosted by NASA in Washington DC.

To find out more about the Conference itself, see:

http://quest.arc.nasa.gov/marsconf/

The focus of the conference was: (as quoted from the NASA web site)

"This summer marks the 20th anniversary of the Mars Viking Landings and 1996 will see new American and Russian missions to the Red Planet. Many educators across the country are gearing up for a special focus on Mars during the upcoming school year.

A three-day teacher workshop to help educators prepare to implement Live From Mars in their classroom occurred in July 18–20 in Washington, D.C.. Optimizing the integration of electronic field trips and the Internet in the classroom was also the focus of the workshop."

> Diversity University will also have a representative at the physical conference to relay our questions to the speakers.
>
> If you do not know what a MOO is, please check out our web site at: http://www.du.org. If you would like an information packet about the MOO with a list of basic commands, contact jeanne@du.org
>
> For those with no web access, DU will provide constantly updating webslate coverage in the MOO of the proceedings which are being transcribed as they occur at the conference site. For those of you who have web access you may access it via the NASA address given above. Following the presentations there will be discussions and during Q & A sessions, our DU representative at the Washington DC site will relay our questions to the presenters.
>
> Be sure to explore the rooms around the main foyer as a number of our students and teachers have created objects, both webbed and non-webbed, that relate to the Mars theme.
>
> DU MOO: (telnet) moo.du.org 8888 (for most systems) (telnet) moo.du.org /port=8888 (for vaxes) (telnet) moo.du.org:8888 (for a few odd birds) NOTE: the port number (8888) is absolutely "required". Otherwise you will reach the logon to our server, rather than the MOO. "

Aye, the MOO instructions do tend to sound like gobbledy-gook to us, and we think inevitably it's going to stay that way for some time. But the technology is growing, no doubt about it.

Just for the fun of it, we've included a shot, as downloaded from Diversity University's publicly available site, of just what virtual reality is like in the student coffee lounge. Looks about as exciting as a real one, to our mind. Hmmmmm.

So there you have it: three sample chat forums or real-time spaces of special interest to scientists, including scientific educators. But you must answer the simple question of your own time, resources and patience: have you got enough of them to schedule yourself into a real time chat on an unwieldy keyboard? Think about it, then decide. Our goal in this book is to provide information for our scientist colleagues to navigate the Internet in practical ways that will promote their science. So frankly, we're not going to promote something with wild enthusiasm that we think could detract from that work. Our recommendation? Proceed with serious caution.

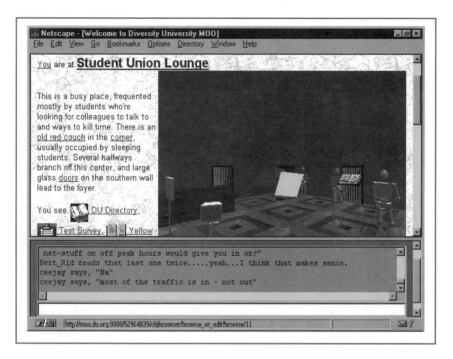

But we don't mean to be seeming to dismiss the serious work that has gone into making this area of the Internet a real growth arena. In fact, we'd like, in these closing comments, to suggest a formalised approach that could enhance the use of these facilities by working scientists who would otherwise wash their hands of the whole affair. It occurs to us that the *de facto* use of IRC, MUDs and MOOs is almost exclusively the Nease Androj's of the scientific world. However, electronic video conferencing of the casual future is going to be built upon the foundations of contemporary IRC systems. So we think that a serious case can be made for deliberate delegation of responsibility for departmental facility in these systems, to the young whipper-snappers coming up through the system, whether sixth-form secondary students, eager-beaver undergrads, graduate students or post-docs, who can then help to train (and even translate for) the grayer heads when the whole discussion system is operative.

What better way to groom your young trainees than consciously to solicit and enhance their netskills so that they will be able to participate in the continued development of global electronic conferencing?

COMMUNICATING WITH OTHER COMPUTERS

File Transfer, or FTP (File Transfer Protocol)

Through her international discussion group, from one of her newer net friends, Jane Q. Sirius has just been alerted to the existence of a Word 6 macro virus that could infect her Mac. All of her current professional life is on that computer, her two grants in preparation, her major review paper, as well as that hard-earned database, and she'd be devastated if it crashed irretrievably. Her contact has given her the Internet address of a special virus detector and eliminator developed specifically for this nasty bug by MicroSoft.

Ooops, but where did she place that letter? Accidentally deleted? Happens to the best of us. Now how can Dr. Sirius find that virus cure? Immediately!

INTRODUCTION

FTP stands for file transfer protocol [in most of the world: KO'D points out that in Scotland's central belt it is shorthand for something else, but then, here in the UK, our popular late night telly show on Friday, that is, 'TFI Friday', also stands for something else]. Anyway, in the previous chapters we saw how files could be sent between individuals. From communicating with other scientists by electronic mail, it's really not a very large step to take to begin to think about getting desirable files yourself. That is to say, instead of waiting passively for files like email or ones that are attached to the email, to find their way to your computer after they've been sent by somebody else, you can actually go out yourself and actively bring them back alive. It's a bit like being able to go to the library yourself and choose a book rather than relying on someone else to do it for you.

In essence you use FTP software to connect to a remote computer, called a 'server' to select a file from its FTP directory and then to bring it back to your own PC, the so-called 'client' in the operation. The type of files available by FTP are obviously less personal than those available by email. If a colleague at a university in the USA wanted your confidential opinion on the draft of a paper she was writing, she wouldn't put it on her university's computer for you to retrieve by FTP, she'd send it by email. The reason for this is that if you could retrieve the file by FTP, in a publicly accessible domain, then so can everybody else who uses the Internet.

The files available by FTP therefore tend to be things that a larger number of people would want to access over time. Examples might be long documents such as the Maastricht treaty, graphic images and especially software.

When you go out yourself into the wild file transfer world, you'll go naturally to the directory on the file server that is specific for your platform .[uh-oh, here we go, more jargon already!] No, what we mean to say is that when you go into FTP mode, you're entering another computer (that is, usually, a publicly available 'server') at a particular directory or folder point, and you need to move into that directory or folder that holds the file you want to acquire. Usually, public servers have files in directories that are platform (ie PC, Mac, Unix) specific. Since you're now safely in your own platform-specific directory, therefore, any troubles opening and using the files you've chosen to download should be minimised. In our experience, getting the files themselves isn't difficult, since the FTP software does that for you. The problems appear only when you want to use the file, as one does, of course, and it's in a strange and unrecognizable format.

A Brief Explanation:

It should not have escaped the astute scientist reader's attention that we have organised this book in a deliberate way; we are moving you from communicating via computer with other human beings in faraway places, via email, newsgroups, and even MUDs and MOOs, if you can bear it, to communicating directly with computers (themselves) in faraway places. Actually, it's all an insidious plot — even now, the conspiracy theorists are organising a grand Luddite wrecking ball. Next thing it'll be just computers talking to computers. But of course, you're always talking to people, through their writing and creating, if at some further remove. Still. We are not yet in that spooky world of Artificial Intelligence.

Seriously, important as it is to be able to communicate with other humanoid creatures busily engaged in troglodyte-ish study on an arcane subject that only boffins can pronounce, let alone understand (do you get the idea from that chip on our shoulder that scientists in the UK might possibly have a less than exalted social standing?) — important as our communication with our peers is, it's also very important in the

Internet world to be able to communicate directly with the computer. The reason for that is that computers can do all sorts of useful things for you that humans can't. They'd be pretty dull participants in a newsgroup where actual thinking was required but they are good at other things. There aren't many people prepared to keep thousands of useful files and software on their PC and send them off on request to thousands of strangers from all over the world. Luckily, there are computers all over the world who are content to carry out that very task.

You've discovered that yourself, sitting at your PC (you don't tickle it right, and it doesn't give you the goods) and so it is with the outside world. There's lots of information and software out there, which you can procure, untouched by human hands except as your own fingers command. This chapter and succeeding ones are concerned with tickling faraway computers so that they disgorge their stored information to you.

Hacking it ain't, because in these cases the computers are publicly accessible; indeed, they're waiting out there for you to access them. No doubt it warms their heart's cockles, and all that. Well, makes them work just that little bit harder, anyway, than if you weren't asking them.

As we write these chapters, the Internet continues to transform. We may have mentioned already that the front end [*the driver's seat, steering wheel and console, as it were, of our simile car*] of your communication entrance to the Internet is rapidly changing, so that individual pieces of software are being assimilated into just the one simple user-friendly console. During this time of transition, it's still important, we think, to understand the individual pieces, because they're still useful as stand-alone components of a great system. They do some useful things that the one-stop shops can't do.

SOFTWARE THAT ENABLES YOUR COMPUTER TO RETRIEVE FILES FROM ANOTHER COMPUTER:

We've grown up on the net over the past couple of years, and so to us these separate functions (i.e. electronic mail, file transfer, and the ones coming below like telnet and World-Wide Web browsers, MOOs and MUDs) have become synonymous with different software packages. You click on one particular package, open it up, and follow the directions to get where you want to go, and procure what you want to retrieve.

However, with the advent of sophisticated WWW browsers, like Netscape Navigator 2.0 and successors, as discussed in the WWW chapter below, these functions are operated apparently seamlessly by the single browser package, which links both retrieval and display software into one unit, or front-end (our car simile was pretty apt, wasn't it?). So by the time this book is published, it's quite likely that we'll all just be using our browsers (check out the glossary on 'browser', if you need to, but essentially a browser is a piece of software that lets you move about on the Internet) to perform these separate jobs. It'll be a case of clicking on the hypertext link (or specifying the address, as discussed below), or the browser button and snap, you're retrieving the file, or checking your post, or looking at that really useful newsgroup; end of story. In the meantime though, it's useful (and might still come in handy) to have some feeling for the trusty old FTP servers, which are not quite so user friendly as WWW pages (discussed below) might seem. FTP software is designed to do one thing and to do it well — to move data from one place to another. That could be from a remote computer to your PC or it could be *from* your PC *to* the remote computer, so that others can access it. That's one important task that browsers won't do (conveniently, yet!).

Also, the individual stand-alone pieces each take up a great deal less operating memory, and so you can stack up various applications, all at once, on your desktop. Quite a convenience, actually. Just because that big browser is newer and flashier, it ain't necessarily better. And our point in this book is to help scientist readers to expedite their way as conveniently as possible, with the minimum of fuss and time spent, through the net.

But we know that you can't hold back the tide of change, just look at 'gopher'. "Say what?" Exactly.

Meanwhile, FTP in whatever incarnation is one of the most important, and at the same time, one of the most infuriating components of the armoury for essaying into the Internet.

BRINGING THE FILE BACK ALIVE:

The basic problem is simple, but can be fiendishly difficult to unravel. Essentially, you've got to have special software installed on your machine, usually but not always provided by your service provider so that you can decompress / expand / unstuff / unzip or otherwise open the file to get it into its normal operating mode. That's because there's no point, from the file server's point of view, in clogging up their memory space with lots of uncompressed files, when they can provide the same service, and save on Internet communication time in the bargain, by providing compressed / stuffed / zipped files. A user-friendly analogy might be the difference between standard and long-play video tapes. You can save a lot of space by recording in long-play mode, but you can't play these tapes back in standard mode. Well, the analogy breaks down a bit, but we hope you see what we're getting at — you have to 'revive' the sleeping file back into standard mode before you can use it on your own machine. We'll be getting into these important expansion utilities in some depth, because you have to have them even if you have the pretty Navigator console, but if you want a sneak preview, check out Table 2 on a page close to you.

In the case of file transfer through an FTP server, it seems it's always the file procurer (as opposed to the sender) who must figure out on their end how to decode / extract / decompress the file they've procured. That way of doing things is just the way the net works. Well, if you can't open a particular file that you've downloaded from the FTP server, you can complain to the poster or to the very busy server monitors, but that would make you look a bit of an ass, wouldn't it, considering the legions of other users who've managed to download successfully. And are you going to get an answer to your plaintive query from the busy computer operators? Think again, chump. If they were that helpful you wouldn't need this book would you?

So, okay then, let's deal specifically with (a) getting that really useful file from an FTP server that somebody from half way around the world mentioned in a mailing list you happen to subscribe to, and (b) ensuring it works successfully on your own personal computer. Probably the first point that we should make, considering the overloading on trans-oceanic cables, is that you should seek to find a 'mirror site' close to home. A mirror site is exactly what it sounds like, only in this case the 'images' (i.e. files) are the same, not mirror images. Confusing to chemists trained to recognise optical isomers, we know, but that's language. If this mirroring bit is our first point, it's so important that we're also going to make it our last, too.

The second important point to make concerns the appropriate address of the FTP server, and the file that you want to procure. Files, as we shall continue to explain, are found in directories on publicly accessible servers, arranged in much the same way as you will find files in your own PC, whether you call the hierarchy 'directories' or 'folders'. File servers are named in what looks like an email address without the '@'. As in, for example, 'src.doc.ic.ac.uk', which is one of our favourites. We'll show you where to put this address, in your file transfer software or application programme, below.

There are only a few really major FTP servers of much use to scientists, and these are mostly concerned, frankly, with crucial bits of software that can help us work on our computer better. These are the sort of shareware sites that the computer magazines raid to pack into their CD-ROMs and flog their mags with. Basically, you might wonder what's the point in forking out the price of the magazine when you can get the files yourself from the servers. Well, we're talking downloading time here, and that's an important consideration during the week, when you're trying to get an important file transferred between the USA and the UK, for example. Weekends are an incredible boon to the Internet, in terms of speed of access and data transmission. But you might want to be thinking about your own home phone bill too. Remember we mentioned tricks with mirrors, right? It's highly likely that the software you want is sitting handily in a 'mirror site' right here in jolly old England. Or in Europe, perhaps. Or is that the same thing? Er, not quite. Oh, we see. The important thing is that you should try to avoid downloading transatlantic data, since these links are already overloaded. It's in your interests to do this since it will save you time and, if you're using your home connection, money. The rule of thumb is: download from the FTP server with the file you want, that's closest to you!

Table 1   List of the Most Useful Anonymous FTP sites in the UK.

ac or Academic		
	ftp.ecs.soton.ac.uk	scott.cogsci.ed.ac.uk
agora.leeds.ac.uk	ftp.ed.ac.uk	src.doc.ic.ac.uk
al.mrc-lmb.cam.ac.uk	ftp.epcc.ed.ac.uk	sunacm.swan.ac.uk
http:www.eia.brad.ac.uk	ftp.kcl.ac.uk	svr-ftp.eng.cam.ac.uk
camelot.cc.rl.ac.uk	ftp.mcc.ac.uk	tardis.ed.ac.uk
catless.ncl.ac.uk	ftp.mechnet.liv.ac.uk	unix.hensa.ac.uk
ccn7.nott.ac.uk	cumulus.met.ed.ac.uk	wombat.doc.ic.ac.uk
cs.ucl.ac.uk	ftp.mrc-apu.cam.ac.uk	ftp.maths.warwick.ac.uk
elvis.sccc.ac.uk	ftp.ncl.ac.uk	
emwac.ed.ac.uk	ftp.ohm.york.ac.uk	co or Company
ftp.brad.ac.uk	ftp.ox.ac.uk	
ftp.brunel.ac.uk	ftp.salford.ac.uk	ftp.acorn.co.uk
ftp.cl.cam.ac.uk	ftp.shef.ac.uk	ftp.almac.co.uk
ftp.comlab.ox.ac.uk	ftp.soton.ac.uk	ftp.cityscape.co.uk
ftp.cs.bham.ac.uk	ftp.stir.ac.uk	ftp.demon.co.uk
ftp.cs.city.ac.uk	ftp.tex.ac.uk	ftp.lasermoon.co.uk
ftp.cs.nott.ac.uk	ftp.warwick.ac.uk	ftp.mantis.co.uk
ftp.cs.york.ac.uk	ftp.york.ac.uk	ftp.micromuse.co.uk
ftp.csc.liv.ac.uk	hcrl.open.ac.uk	ftp.net-shopper.co.uk
ftp.dai.ed.ac.uk	julius.cs.qub.ac.uk	ftp.pavilion.co.uk
ftp.dcs.ed.ac.uk	kuso.shef.ac.uk	ftp.tecc.co.uk
ftp.dcs.gla.ac.uk	lister.cc.ic.ac.uk	web.nexor.co.uk
ftp.dcs.kcl.ac.uk	micros.hensa.ac.uk	http:www.rednet.co.uk
ftp.dmu.ac.uk	mscmga.ms.ic.ac.uk	
ftp.doc.ic.ac.uk	newton.newton.cam.ac.uk	org or Organisation
ftp.dur.ac.uk	nyquist.cs.nott.ac.uk	
	scitsc.wlv.ac.uk	ftp.bbcnc.org.uk
		ftp.wcmc.org.uk

These FTP servers can also usually be accessed using your WWW browser, by using the prefix 'ftp://' before the name as specified above. For example: 'ftp://ftp.doc.ic.ac.uk' keyed into the 'Open URL' window of your browser should bring you to the same site and directory as keying in 'ftp.doc.ic.ac.uk' does on a dedicated FTP application.

See also, on the World Wide Web: http:// tile.net /ftp-list/ for a full listing of ftp servers by country, as currently specified below

Antarctica	Hong Kong	Portugal
Argentina	Hungary	Romania
Australia	Iceland	Russia
Austria	India	Singapore
Belgium	Indonesia	Slovakia
Brazil	Iran	Slovenia
Byelorussia	Ireland	South Africa
Canada	Israel	South Korea
Chile	Italy	Spain
Chili	Japan	Sweden
China	Korea	Switzerland
Colombia	Latvia	Taiwan
Croatia	Lithuania	Thailand
Cyprus	Luxemburg	The Czech Republic, Prague
Czech Republic	Malaysia	The Netherlands
Denmark	Mexico	Turkey
Ecuador	Netherlands	UK
Finland	New Zealand	Unknown
France	Norway	US
Germany	Peru	USA
Germany USA	Phillipines	Venezuela
Greece	Poland	

We were speaking about getting answers for a difficult FTP problem, and it's certainly nice to know that other people might have similar questions. So it could be worth a wee jaunt into the Usenet Newsgroup (we've discussed newsgroups in Chapter 3, right?) called 'comp.answers'. In this newsgroup, as a matter of compelling interest, answers to FAQs are regularly posted, including exhaustive lists of public FTP servers around the world, specific to particular platforms. Even Acorn platforms get a wee mention on a fortnightly basis!

FINDING NICE SOFTWARE THAT YOU CAN USE IN YOUR DAY-TO-DAY WORK

But let's suppose, just suppose, a friendly colleague in a friendly newsgroup, or in your useful subscription list, tells you the address of an incredibly useful piece of software (or even a useful database file — let's not forget that some important databases are out there too). You want to go out and get it, with a minimum of fuss.

Model of How to Go About Getting Something Important From a Specified or Even From an Unspecified FTP Site:

As an example, let's consider the very useful 'Weather or not' package we recently saw announced in the K12.education.science newsgroup (Refer back to the Newsgroup chapter for further information on newsgroup access and importance). The person making the announcement casually noted that the package, which holds lots of good information on weather monitoring, including lesson plans and interactive type stuff ideal for the classroom, is available on the "standard sumex-aim mirrors."
Eh?

In my youth (when we started this book — seems like a long time ago now) I didn't know where to look for the mirrors, so I emailed the poster of the original announcement, saying, basically, "Eh?" and he kindly forwarded along an exhaustive list of the sites in Europe which mirror the main 'sumex-aim' [and before you ask, we don't exactly know just what 'sumex-aim' is, either!] offerings at the University of Stanford site in California. Most of these sites are updated daily. We don't think it's worthwhile to include this list here, as we've indicated other ways to search for files below which are just as convenient, and which will cover the entire list anyway. But it was nice to get a direct response back from the person who was promoting the application. Anyway, since we're in the UK, I looked for a mirror site in the UK, and sure enough, there was the mirror at the 'src.doc.ic.ac.uk' FTP site. Good old Faithful! This example is another indication of the general utility of the Internet: if you don't know something, and (heaven forfend) it's not in this book, don't be (too) afraid to ask! Direct your query, if possible, however, to the nearest direct person to the answer, as I did above.

So this model should be a good demonstration of the powers of newsgroups, of email, and of appropriate FTP retrieval, as well as the worthwhile wonders and time-saving effects of the Internet, innit?

Now on old software dedicated to FTP (Like Fetch, or Anarchie, as discussed below), you'll key in the address of the FTP server (src.doc.ic.ac.uk), and put 'anonymous' as your name, and perhaps (probably) your email address as password (some servers like this, some don't care, and some like to have a '+' inserted before the password proper, so you can get files in unzipped format, if that's your pleasure, but we're getting ahead of ourselves again). The FTP software screen may look something like this:

```
╔════════════════════════════════════════════════════╗
║ ▦▦▦▦▦▦▦▦▦▦  Open Connection...  ▦▦▦▦▦▦▦▦▦▦ ║
║                                                      ║
║  Enter host name, user name, and password            ║
║  (or choose from the shortcut menu):                 ║
║                                                      ║
║   Host:       │ src.doc.ic.ac.uk              │      ║
║                                                      ║
║   User ID:    │ anonymous                     │      ║
║                                                      ║
║   Password:   │                               │      ║
║                                                      ║
║   Directory:  │                               │      ║
║                                                      ║
║   Shortcuts:  │▼│    ( Cancel )     ( OK )            ║
║                                                      ║
╚════════════════════════════════════════════════════╝
```

Then you're usually given an opportunity to specify which directory you want to look at, on the FTP server, in a standard format, (like /packages/info-mac/) This specification is quite a time-convenience, because it saves you the trouble of clicking through all the directories once you arrive at the server. But for now, let's just get into the main gate — ready now, click on the OK button, and presto, we're whisked into the Imperial College file server.

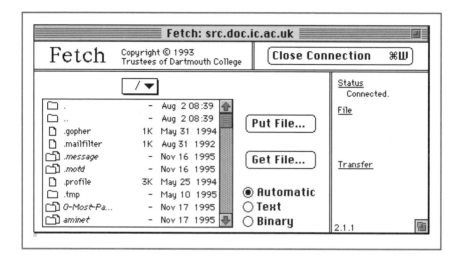

Okay, suddenly, we're 'in' the main directory of the src.doc.ic.ac.uk FTP
server. There's a lot of folders here, arranged alphabetically, (though there
may be several apparent alphabets!) and we want to get down to the
'packages' folder. Just move down with the arrow, or it may be convenient
to specify the letter 'p' directly from your keyboard. Here we are:

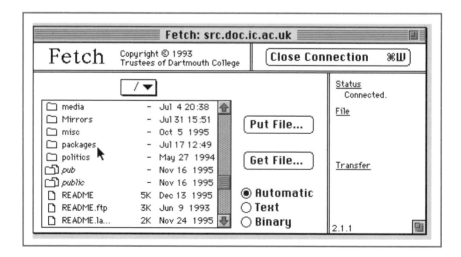

You can easily move around directories by clicking and double-clicking,
until you find the file you want. In this case, if we open the 'packages'

folder, we find lots of packages, and we scroll down again until we reach the 'info-mac' one:

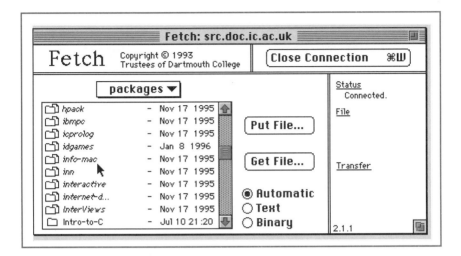

Yikes! Being a methodical, reasonable sort of scientist, I've already arrowed down in the info-mac folder until I reached the S's. Now it's a simple matter to double-click on the 'Sci' folder, and within a couple of seconds I see the titles of several hundred files. You can tell they're files since they look like a sheet of paper in portrait mode, instead of the folder icon. And moving down with the handy arrow, or again, prompting the ftp server with the key 'w' I come to the 'weather or not' file that I wish.

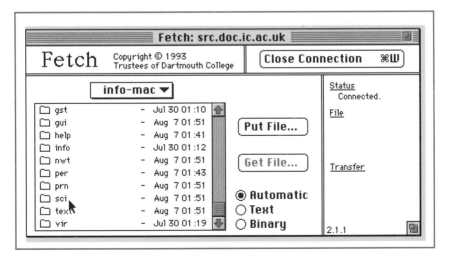

You might find that it's conveniently sort of obscure, but you can see the whole name simply by increasing the size of the Fetch window, whereupon all becomes clear:

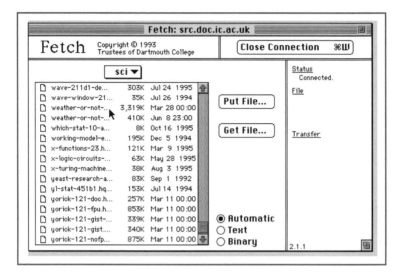

It might be worthwhile just noting the directory route, here in helpful graphics mode. We've come in through the main directory, denoted by the slash mark, through the packages and info-mac directories into the Sci directory, and here we've found the weather-or-not file we wish to retrieve.

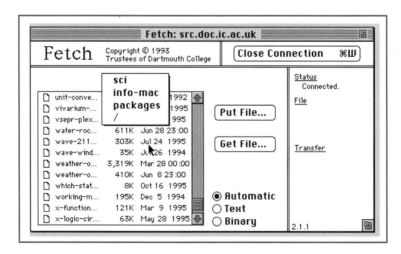

When you've identified the file you want, file transfer will be a simple matter of specifying that file (whether by double clicking, or highlighting and pressing the 'Get 'button). The Fetch software shows this in red, and the file name is still legible.

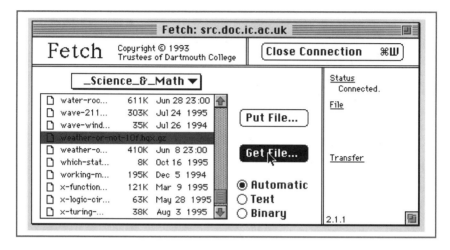

and then specifying in which directory on your own machine you want to save the retrieved file to. You'll be prompted as to how on your own machine you want to save the file, as so:

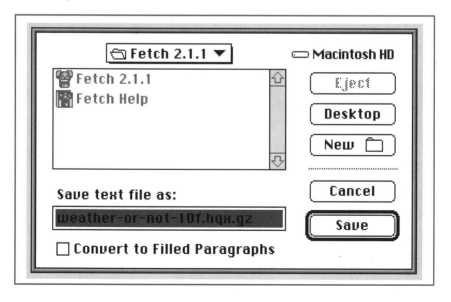

Then you sit back and wait. The rate of transmission is usually specified by the software, so you can figure out how long, roughly, it's going to take to download.

Here's an example of the Fetch software in action:

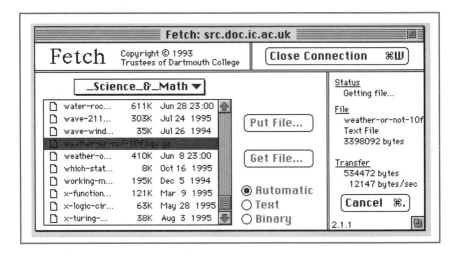

It's cranking in at the moment at the relatively slow rate of 12 kb/sec, which on a file of 3.3 megabytes will take roughly oh, 300 seconds? More or less anyway.

File transfers are often a good time to have a cup of coffee, or tea. And why not? You've been slaving over the computer console for at least 3 minutes getting into the site, finding the right directory, and hey, you deserve a break. You'll be able to communicate with those other humanoids who share your building.

* * * * *

Well, here we are again (back in the present time). It turns out that in fact I've retrieved this 3.3 meg file in about 1 and a half minutes, averaging around 40kb per second. Good-bye coffee break. Well, that's on the ethernet through the university connection, at 1:00pm, a time which is pretty typical of UK lunch times, and Newcastle is connected to the IC server via SuperJanet, so download times are pretty damn impressive.

Just to re-cap then, remember that the 'Weather or not' file's name together with the FTP address, looks like this:

src.doc.ic.ac.uk:packages/info-mac/sci/weather-or-not-10f.hqx.gz

The ftp server is specified, and the directories and file name are shown after the colon, so that you can identify to your FTP software where you want to go.

Sometimes the file might be specified by the convention

ftp://src.doc.ic.ac.uk/packages/info-mac/sci/weather-or-not-10f.hqx.gz

This specification, or Uniform Resource Locator, or 'URL' is specific for World Wide Web browsers, and more about them in a later chapter, but we thought that it might be convenient for you to know and understand the difference in the two designations.

We'd just like to note, however, if you're using a contemporary browser like Netscape (and isn't everybody?) that merely specifying that URL on the user-friendly console will also immediately initiate the file transfer. Netscape places its retrieved files by default into the System Folder, in case you can't figure out where it's gone once it's been retrieved, but you can change the destination by choosing the options and preferences in the Netscape menus; some people do like to be able to specify into which folder their newly acquired material should go. By the way, increasingly Netscape, if you or your system operators have installed the appropriate bits of software, and directed it well in setting it up, will determine the appropriate action to be taken with retrieved files, so that all this following unzipping and unstuffing will become unnecessary, having been done by default. [And a good thing, too!]

GOT MY FILE, CAN'T OPEN IT!

Anyway, let's assume you've got a file, and it's sitting there and no application will open it, and you're getting damned frustrated! It's probably important to note then whether this particular file is 'zipped' (i.e. .gz) as well as 'stuffed' (.hqx). Therefore once I retrieve it, I will want my software to unzip and unstuff it, before I can use it. Usually Windows users only have to unzip their files, which they've retrieved as .zip files, but then that's fair, as they're usually a bit more challenged than [smug] Apple users. Often virus checkers (check out the Tea Break chapter below for a bit more on security) don't cope with zipped or stuffed files, so you have to get them in standard mode before you can do a proper hygiene check on them.

So now it's time to look closely at that table on File Compression/ Expansion Utilities, ah that'll be Table 2. Choose your utility according to your platform, and needs, download from the nearest available server if you haven't got such useful software devices already, and begin to unzip, or indulge in a bit of how's your father. On Macs, at least, applications called say, 'Stuffit' can also 'unstuff' so that can be a bit confusing [*Well, it was pretty confusing to me!* — LW]. You have to see what the application can do, in your individual case.

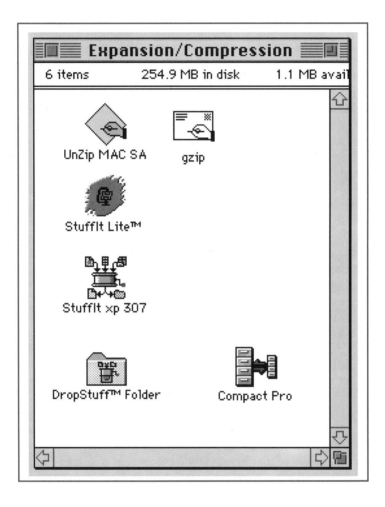

Table 2   Useful Utilities (Expansion and Compression) that you may not yet have

Table 2a   Windows-Based Compression/Expansion Software

PKUnzip

achilles.doc.ic.ac.uk:/imported/oxford/techpapers/Ian.Page/sundance/pkunzip.exe
unix.hensa.ac.uk:/mirrors/Hyper-G/Easy/pkunzip.exe
unix.hensa.ac.uk:/mirrors/Hyper-G/pc-client/amadeus.100/pkunzip.exe
unix.hensa.ac.uk:/mirrors/matlab/books/stonick/pc/pkunzip.exe
unix.hensa.ac.uk:/mirrors/statlib/DOS/dos-tools/pkunzip.exe

WinZip

micros.hensa.ac.uk:/mirrors/simtelnet/win95/compress/winzip95.exe
micros.hensa.ac.uk:/mirrors/uthscsa/utils/winzip95.exe
micros.hensa.ac.uk:/mirrors/winsite/win95/miscutil/winzip95.exe
micros.hensa.ac.uk:/micros/ibmpc/win32/a/a069/winzip95.exe
ftp.hea.ie:/part5/win95/miscutil/winzip95.exe
ftp.hea.ie:/part6/simtel-nt/archiver/winzip95.exe
ftp.hea.ie:/part6/simtel-win95/archiver/winzip95.exe
warum.uni-mannheim.de:/systems/windows/win32/win95-cica/miscutil/winzip95.exe
warum.uni-mannheim.de:/systems/windows/win32/win95-simtel/compress/
winzip95.exe
warum.uni-mannheim.de:
      /systems/windows/win32/win95winsock/Misc_Utils/Compression/
winzip95.exe

Table 2b   Mac-Based Utilities Software

MacGzip

ftp.loria.fr:/pub/mac/umich/util/compression/macgzip1.0b0.sit.hqx

nic.switch.ch:/mirror/umich-mac/util/compression/macgzip1.0b0.sit.hqx

sunsite.rediris.es:
/software/umich-mac/util/compression/macgzip1.0b0.sit.hqx

ftp.hrz.uni-kassel.de:
/pub4/mac/mac.archive.umich.edu/util/compression/macgzip1.0b0.sit.hqx

Stuffit Expander

DropStuff with Expander Enhancements and Stuffit Expander are an excellent joint package for expanding virtually any compressed Mac file

jeeves.ncl.ac.uk/temporaryftp/utilities/ both of the above (v 4.0) are available at the jeeves.ncl site

unix.hensa.ac.uk:/mirrors/uunet/networking/info-service/gopher/Macintosh-TurboGopher/helper-applications/StuffItExpander_3.5.2.hqx

nic.switch.ch:/mirror/gopher/Macintosh-TurboGopher/helper-applications/
StuffItExpander_3.5.2.hqx

Table 2b   continued

Compact Pro

unix.hensa.ac.uk:/mirrors/uunet/networking/info-service/gopher/Macintosh-TurboGopher/helper-applications/CompactPro1.50.hqx

sun.dante.de:/tex-archive/tools/gopher/Macintosh-TurboGopher/helper-applications/CompactPro1.50.hqx

nic.switch.ch:/mirror/gopher/Macintosh-TurboGopher/helper-applications/CompactPro1.50.hqx

fsuj01.rz.uni-jena.de:/pub/tex/archive-tools/gopher/Macintosh-TurboGopher/helper-applications/CompactPro1.50.hqx

fsuj01.rz.uni-jena.de:/pub/tex/tools/gopher/Macintosh-TurboGopher/helper-applications/CompactPro1.50.hqx

ftp.ask.uni-karlsruhe.de:/pub/infosystems/gopher/Macintosh-TurboGopher/helper-applications/CompactPro1.50.hqx

ftp.univie.ac.at:/packages/network/gopher/Macintosh-TurboGopher/helper-applications/CompactPro1.50.hqx

[The following is a semi-confusing bit of real time unstuffing, just to give you the impression of what happens when the unstuffing or unzipping sequence swings into action; we recommend humming the Benny Hill theme tune (AKA 'Yakety Sax') while you read it.]

Now I simply double-click on the zipped/stuffed file, and my Gunzip software swings into action. Once the file is unzipped, the software gives me a pretty sounding 'Ding!' and I note that its name has changed from "weather-or-not-10f.hqx.gz" to "weather-or-not-10f.hqx". This Gunzip programme doesn't bother to leave the old 'zipped' file, but just replaces it with the unzipped one, which is now 4.4 meg. The hqx suffix means that I will need to use an unstuffing or expanding package; I can drag the hqx file onto the application program (*hey, I'm an apple man, I pull it by its hair!* — LW), and watch while it does its work. In a moment or so, I see three files: the original hqx file, a new one called a .sit file, and a whole new folder (called Weather 1.0f) in which I can find 3 interesting files. These files are: 'Read me first', the Weather application, and something else, a movie document. Apparently the .sit file has extracted itself into the new folder, so I delete it as redundant, as well as the hqx file — I have a very limited hard drive, and need to conserve space as much as possible!

In order to look at these expanded files, finally, of course I have to have the software in my PC to access them: a simple text application to read the 'Read me first' document, and a movie-processing application to see the movie. Double clicking on the 'Read me first' document elicits my SimpleText application, and I'm suddenly reading the instructions for use. The Weather file is itself an application, and now that I've read the 'Read me first' documentation, and my machine is configured properly, and I've followed all appropriate installation instructions, then it should fly! In this particular installation, I only needed to ensure that the preferred 2 meg RAM is allocated to the application (it was), as noted carefully in the instructions.

By the by, it will only have taken me a few moments to have checked the unstuffed package for any nasty viruses, using one of the antiseptic applications we note in a subsequent chapter, right? Right.

So now it's on to double-clicking on the Weather application, and seeing whether this wonderful program can help teach struggling secondary school students about weather.

'Scuse me while I just take a wee peek.

The take-home message here is that it does look like a lot of fun! So more power and thanks to the originator. This seems like as good a time as any to discuss the concepts of freeware and shareware, as distributed on FTP servers.

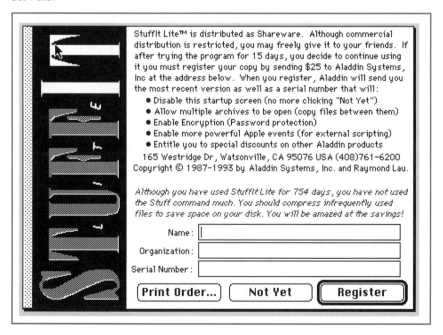

StuffIt Lite™ is distributed as Shareware. Although commercial distribution is restricted, you may freely give it to your friends. If after trying the program for 15 days, you decide to continue using it you must register your copy by sending $25 to Aladdin Systems, Inc at the address below. When you register, Aladdin will send you the most recent version as well as a serial number that will:
- Disable this startup screen (no more clicking "Not Yet")
- Allow multiple archives to be open (copy files between them)
- Enable Encryption (Password protection)
- Enable more powerful Apple events (for external scripting)
- Entitle you to special discounts on other Aladdin products

165 Westridge Dr, Watsonville, CA 95076 USA (408)761-6200
Copyright © 1987-1993 by Aladdin Systems, Inc. and Raymond Lau.

Although you have used StuffIt Lite for 754 days, you have not used the Stuff command much. You should compress infrequently used files to save space on your disk. You will be amazed at the savings!

Name : |

Organization :

Serial Number :

[Print Order...] [Not Yet] [Register]

A NOTE ABOUT FREEWARE AND SHAREWARE:

We're not computer programmers, though sometimes we wish we were, considering the vast salaries the good ones (and even not so good ones) seem to get. One of the ways, apparently, that computer programmers can make their (free-lance) living, again if they're good, is by creating a useful programme that everybody just has to have. Examples: the unstuffing software we've been discussing, above and below. These shareware packages often come with a price tag, a nominal amount, really, of some $15–$25.00 which, if you do use the programme a lot, you are really morally obliged to pay, aren't you? If you don't pay, you get these little flashing indicators (nag screens) at the start of the application, to wit: 'Do you wish to register this copy of XXXX?' And then you send your money along by cheque to the person whose address is given in the flashing sign. Now just pretend for a moment that say 100,000 people are using your incredibly useful software, and forwarding along a nominal $10 on to you. Good money by any body's standard, eh? So the proposition becomes worthwhile, even if 50% of the people who gladly use the shareware can't be bothered (or are too poor) to return the registration fee. And that's how shareware programming continues to be an important component of the information revolution. Think of it as a sort of electronic honesty box.

The real freeware is just that, absolutely and unequivocally free, and you don't pay a penny for the privilege of using it. Not a lot more to be said about that, except 'Thank you' and I'm sure donations, nevertheless, are still gratefully accepted if you can figure out who to send them along to!

Sometimes people do things just for the love of it — but then, as scientists, we understand that, don't we?

As we may have mentioned above, as file procurer, you yourself are responsible for opening the file you've retrieved, and the files might come in various compressed formats. A file can be compacted, or stuffed, and zipped after that, so you might have to go through several opening procedures. I myself (LW) have to confess to having had difficulty with unzipping functions in the past, though I think I've got it sussed now; I'm afraid this is a trial and error (or as we say in the north, suck it and see!) scenario. If you experience horrible difficulty, and are pulling your hair out, that's the time to ask for advice, either of your Internet Service Provider, or in other forums, like newsgroups dedicated to questions, or platform-specific subscription lists, or even us!

So it's an almost inevitable fact of life (like pulling teeth) that you've got to have the unzipping, uncompacting, unstuffing software on your machine before your files can be brought back to normal, working size, although self-extracting files are very convenient too. This is true whether or not you've got the Netscape WWW browser, because even Netscape has to be pointed to an application that fits the appropriate file suffix. This is all beginning to sound like a Catch-22, or the great non sequitur, because how can you unstuff without an unstuffer? Especially if the unstuffer is stuffed? The great thing about the FTP servers is that you can get the appropriate software required to do these functions directly from the FTP servers themselves — in unzipped, unstuffed mode, of course, or even in self-expanding mode. Yes, sometimes you get files that expand all by themselves! However, inevitably your work will only proceed in something of a bootstrap fashion until you have all the kit you need. Thankfully, these incredibly useful and timesaving bits of software are frequently available through your own service provider, but if not, you can get them, in self-expanding or already expanded form, from the list of addresses and directory sections we've included already, in Table 2. FTP software will probably be part of the basic toolkit supplied by your ISP, but otherwise, we hope we've helped you through this very frustrating part of the Internet jungle.

SEARCHING FOR THAT FILE:

One more bit of advice before we move on to a pertinent comment on a really useful FTP search device. We've noted already that downloading files, especially on cross-Atlantic connections, can be particularly time consuming (and often fruitless, as you wait for those dribs and drabs of the file to flick across, until finally you're forced to cut your losses, and hang up). Before you even get to such a frustrating point, when you find yourself shaking your fist and cursing at the screen, it can be worth while checking things out with your favourite server in your local area. If you have to, spend a wee bit of time clicking through those directories — it will still beat all hell out of waiting hours for a trans-Atlantic transmission, which might not get completed anyway. Speed of transmission inside the country is really dramatically enhanced, as compared to the speed of packet exchange particularly on trans-Atlantic routes. And why clog up the system unnecessarily, anyway?

And now for the pertinent comment. How can you find out easily whether the file you want is actually sitting smugly in an FTP server almost 'in-house' as it were? Here's where we all have to take our hats off to the shareware creator of 'Archie' or 'Anarchie', the FTP searcher's companion!

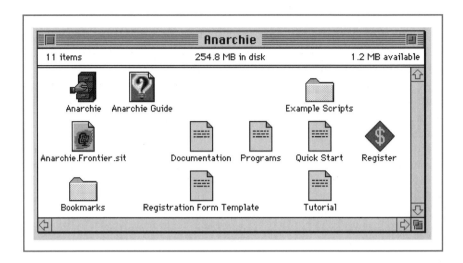

You can find this software in our useful Table 3 along with all FTP addresses for all the other components of crucial Internet access software. Archie will take a query from you (say a string of characters) and query the other FTP servers around the world for that string. Archie has an internal set of 'bookmarks' or servers which it queries for a particular keystring. An example of part of the bookmark set is shown below.

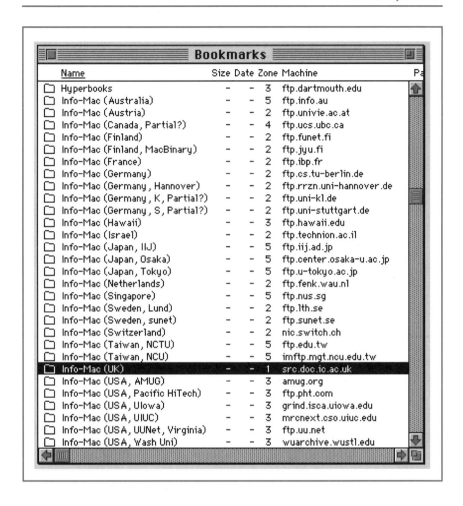

Name	Size	Date	Zone	Machine	Pa
☐ Hyperbooks	–	–	3	ftp.dartmouth.edu	
☐ Info-Mac (Australia)	–	–	5	ftp.info.au	
☐ Info-Mac (Austria)	–	–	2	ftp.univie.ac.at	
☐ Info-Mac (Canada, Partial?)	–	–	4	ftp.ucs.ubc.ca	
☐ Info-Mac (Finland)	–	–	2	ftp.funet.fi	
☐ Info-Mac (Finland, MacBinary)	–	–	2	ftp.jyu.fi	
☐ Info-Mac (France)	–	–	2	ftp.ibp.fr	
☐ Info-Mac (Germany)	–	–	2	ftp.cs.tu-berlin.de	
☐ Info-Mac (Germany, Hannover)	–	–	2	ftp.rrzn.uni-hannover.de	
☐ Info-Mac (Germany, K, Partial?)	–	–	2	ftp.uni-kl.de	
☐ Info-Mac (Germany, S, Partial?)	–	–	2	ftp.uni-stuttgart.de	
☐ Info-Mac (Hawaii)	–	–	3	ftp.hawaii.edu	
☐ Info-Mac (Israel)	–	–	2	ftp.technion.ac.il	
☐ Info-Mac (Japan, IIJ)	–	–	5	ftp.iij.ad.jp	
☐ Info-Mac (Japan, Osaka)	–	–	5	ftp.center.osaka-u.ac.jp	
☐ Info-Mac (Japan, Tokyo)	–	–	5	ftp.u-tokyo.ac.jp	
☐ Info-Mac (Netherlands)	–	–	2	ftp.fenk.wau.nl	
☐ Info-Mac (Singapore)	–	–	5	ftp.nus.sg	
☐ Info-Mac (Sweden, Lund)	–	–	2	ftp.lth.se	
☐ Info-Mac (Sweden, sunet)	–	–	2	ftp.sunet.se	
☐ Info-Mac (Switzerland)	–	–	2	nic.switch.ch	
☐ Info-Mac (Taiwan, NCTU)	–	–	5	ftp.edu.tw	
☐ Info-Mac (Taiwan, NCU)	–	–	5	imftp.mgt.ncu.edu.tw	
☐ Info-Mac (UK)	–	–	1	src.doc.ic.ac.uk	
☐ Info-Mac (USA, AMUG)	–	–	3	amug.org	
☐ Info-Mac (USA, Pacific HiTech)	–	–	3	ftp.pht.com	
☐ Info-Mac (USA, UIowa)	–	–	3	grind.isca.uiowa.edu	
☐ Info-Mac (USA, UIUC)	–	–	3	mrcnext.cso.uiuc.edu	
☐ Info-Mac (USA, UUNet, Virginia)	–	–	3	ftp.uu.net	
☐ Info-Mac (USA, Wash Uni)	–	–	3	wuarchive.wustl.edu	

A similar bookmark set should come with your own copy of Anarchie. Typically, Archie will query those servers closest to your machine.

So, as an example, let's look for 'weather', and since we're in the UK, we'll ask Archie at src.doc.ic.ac.uk.

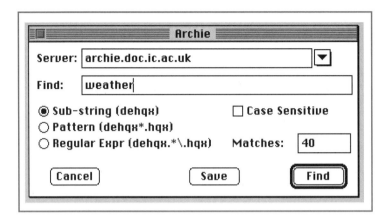

Archie mounts a convenient window to show you that it's working on your request:

Uh-oh, back by return minutes, or seconds depending on current usage constraints,

come the names, paths and directories, as well as FTP sites, of 40 different weather files, and none of them are the one we want!

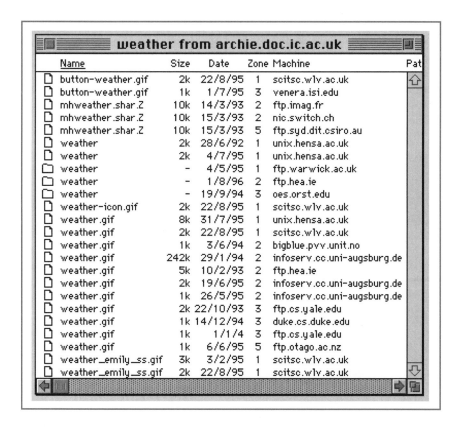

Name	Size	Date	Zone	Machine	Pat
button-weather.gif	2k	22/8/95	1	scitsc.wlv.ac.uk	
button-weather.gif	1k	1/7/95	3	venera.isi.edu	
mhweather.shar.Z	10k	14/3/93	2	ftp.imag.fr	
mhweather.shar.Z	10k	15/3/93	2	nic.switch.ch	
mhweather.shar.Z	10k	15/3/93	5	ftp.syd.dit.csiro.au	
weather	2k	28/6/92	1	unix.hensa.ac.uk	
weather	2k	4/7/95	1	unix.hensa.ac.uk	
weather	-	4/5/95	1	ftp.warwick.ac.uk	
weather	-	1/8/96	2	ftp.hea.ie	
weather	-	19/9/94	3	oes.orst.edu	
weather-icon.gif	2k	22/8/95	1	scitsc.wlv.ac.uk	
weather.gif	8k	31/7/95	1	unix.hensa.ac.uk	
weather.gif	2k	22/8/95	1	scitsc.wlv.ac.uk	
weather.gif	1k	3/6/94	2	bigblue.pvv.unit.no	
weather.gif	242k	29/1/94	2	infoserv.cc.uni-augsburg.de	
weather.gif	5k	10/2/93	2	ftp.hea.ie	
weather.gif	2k	19/6/95	2	infoserv.cc.uni-augsburg.de	
weather.gif	1k	26/5/95	2	infoserv.cc.uni-augsburg.de	
weather.gif	2k	22/10/93	3	ftp.cs.yale.edu	
weather.gif	1k	14/12/94	3	duke.cs.duke.edu	
weather.gif	1k	1/1/4	3	ftp.cs.yale.edu	
weather.gif	1k	6/6/95	5	ftp.otago.ac.nz	
weather_emily_ss.gif	3k	3/2/95	1	scitsc.wlv.ac.uk	
weather_emily_ss.gif	2k	22/8/95	1	scitsc.wlv.ac.uk	

weather from archie.doc.ic.ac.uk

So let's narrow it down a bit; let's see, the announcement said the file was called 'Weather-or-not', so let's specify that string. Back in about 3 minutes comes a wee list of 3 FTP sites (one in Eire, one in Spain, one in Italy) which have mirrored the file. Each site supplies the hqx stuffed form (4.4 meg), but not the zipped form (3.6 meg), which we already know is supplied on the src.doc.ic.ac.uk server. Strange, let's ask Archie in Spain. Back in 3 minutes comes information on the mirrored file in Norway, Sweden, Spain, Switzerland, Italy, Ireland, France, and Germany. What gives? Why haven't we found it on our favourite server?

It's a mystery. An independent query of the src.doc.ic.ac.uk's service indicates that they've been having a wee spot of hardware failure, which could explain the problem with the queries. It happens! But here's the important thing — although we might have missed the 'Weather-or-not' file in src.doc.ic.ac.uk to begin with, we'd still have had our choice of several European sites from which to download the file. And that's got to be better than downloading from Stanford! So a hit and a near miss, on the model, so far. Hey, even the Internet, even this book, ain't perfect. So don't get frustrated if things don't work the way you expect them to, first time. There's almost always another way to get around the problem.

Let's see how we could find it using another very useful FTP search engine, this one accessible using the World Wide Web. For further information on the World Wide Web, see the relevant chapter below. Meanwhile, do consider this very useful FTP search engine in Norway:

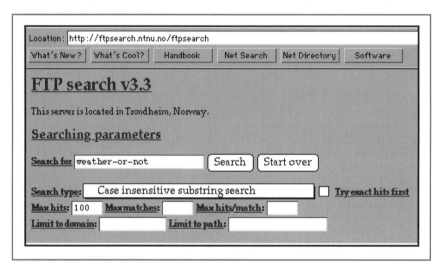

Here we find our lovely 'Weather-or-not' file in a matter of a few moments, and out of the 100 hits, there's familiar old 'doc.ic.ac.uk'. Using your WWW browser, now, you only have to double click on the hotlink, and suddenly you're FTP-ing. Ain't it great!

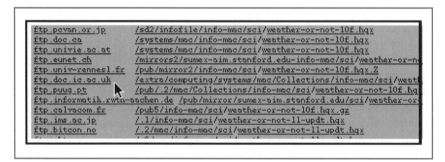

Sorry we had to truncate the last bit of the file name, to fit the figure on the page.

Another way to find information, as we have already suggested, is to check out the newsgroup or mailing list which particularly suits the field or query, but if you're incredibly frustrated you'll probably want to contact a real person, or, for example, email us. But of course, we're struggling scientists and busy people like you, so we probably won't be able to give you immediate attention. Most people like to figure out their own solutions, anyway, if they only knew where to look! And those kinds of answers are all around too.

For example, another very enterprising solution, available on the WWW, to the problem of searching for files on publicly accessible FTP sites has been developed by the folks who can be found at: the URL:

http://www.tile.net/tile/ftp-list/ftpkclacuk.html

who have used the 'Excite' search engine to pose questions to a database of virtually all the world's important 'anonymous FTP servers'. So for example, if, like me, you wanted to find where in the UK you might find the popular freeware for off-line mailing called 'Pegasus', you can key in a few simple words, like 'Pegasus mailer uk' and presto, back comes a list of potentially useful sites, like the FTP site at Bradford, for example, which is pretty close to Newcastle upon Tyne, by the way, so I might as well save on bandwidth and use a really close server. And so that's what I've done in the Pegasus section in Table 3.

PLACING YOUR OWN FILES ON AN FTP SERVER FOR PUBLIC OR SEMI-PUBLIC ACCESS:

In theory, you can also provide access to your own files to other members of your community, if your service provider provides an FTP server (as for example, ftp.demon.co.uk, or say the big university FTP sites like src.doc.ic.ac.uk). Subject to the server's rules and regulations, and the perceived importance of your precious files, you can put files of your own up so that they can be retrieved by other interested parties. In practice, however, unless you've got some incredibly useful files, software, databases, or helpful information whose distribution can be maximised in this way, it's just as easy to attach to individual email and click them off and out. If your WWW pages are on a remote computer though (as they are likely to be) then you will need to use FTP to add and edit them.

If the option of placing your files on your service provider's FTP server does exist, of course, then it would make some sense to do so, if you want to distribute them easily. (and minimise hands-on fuss on your part, once the file is placed properly according to your server's instructions). You can check to see that downloading works, by downloading it yourself. And of course, it's been the downloading that we've really been concerned with in this chapter.

News on the Windows 95 front is that you can use a useful bit of FTP software to make your Windoze into an FTP server to serve yourself. Of course, Apple folks have been able to do this sort of thing, via AppleShare on a network, for some time. We think we'll leave information on setting up the AppleShare system or the Windows 95 serving system to be developed between individual users and their own network managers. Access to your own server, ie your own computer, from outside can be controlled by password entry. You might like to consider questions of security, of course, in a subsequent chapter below.

CONCLUSION:

One of the most important components of the Internet is the sharing of information, [*some would say that information sharing is THE point*] and all of the Internet's tools must serve to facilitate that exchange, and then die, or get transformed into another form, as some better facilitator comes along. 'Ere, we didn't train as philosophers, it just came naturally!'

Table 3   **Internet Software that is described throughout the text**

Table 3a   **PC-Based Internet Software**

Windows95 Internet Software at sunsite.doc.ic.ac.uk

FTP Software
ftp://sunsite.doc.ic.ac.uk//packages/ibmpc/windows95/netutil/cuteftp32.zip
ftp://sunsite.doc.ic.ac.uk//packages/ibmpc/windows95/netutil/ftpic22b.zip
ftp://sunsite.doc.ic.ac.uk//packages/ibmpc/windows95/netutil/ftpx0025.zip
ftp://sunsite.doc.ic.ac.uk//packages/ibmpc/windows95/netutil/ws_ftp32.zip

Telnet Software
ftp://sunsite.doc.ic.ac.uk//packages/ibmpc/windows95/netutil/wintel32.zip

Email Software
ftp://sunsite.doc.ic.ac.uk//packages/ibmpc/windows95/netutil/emailr32.zip

Newsgroup Software
ftp://sunsite.doc.ic.ac.uk//packages/ibmpc/windows95/netutil/newsx86.zip
ftp://sunsite.doc.ic.ac.uk//packages/ibmpc/windows95/netutil/winvn_99.zip

Table 3a **continued**

Windows 3 Software

Archie for Windows:
ftp://ftp.csusm.edu/pub/dosworld/archiezip
http://www.europeonline.com/intl/custom/uk/down/rfudl190.htm
http://volans.co.ul/tucows/files/wsarch10.zip

FTP simple software
ftp://sunsite.doc.ic.ac.uk//packages/ibmpc/windows3/winsock/winftp.zip
ftp://sunsite.doc.ic.ac.uk//packages/ibmpc/windows3/winsock/wftpd202.zip

Telnet
ftp://sunsite.doc.ic.ac.uk//packages/ibmpc/windows3/winsock/wintelb3.zip
ftp://sunsite.doc.ic.ac.uk//packages/ibmpc/windows3/winsock/telftp16.zip

(Off-line mailers:

EudoraPC

warum.uni-mannheim.de:/packages/wais/PC/Dir-Eudora/eudora14.exe
ftp.cnr.it:/pub/PC-IBM/umich.edu/communications/winsock/eudora14.exe
idea.sec.dsi.unimi.it:/pub/security/crypt/rpub.cl.msu.edu/pc/win/winsock/
eudora14.exe
teseo.unipg.it:/pub/win3/winsock/eudora14.exe
ftp.hrz.uni-kassel.de:/pub1/net/mail/popmail/ms-win/eudora14.exe
ftp.ask.uni-karlsruhe.de:/pub/infosystems/gopher/ms-windows/apps/eudora14.exe
ftp.dfn.de:/pub/net/pc-tcp-ip/winsock_cica/eudora14.exe
ftp.ms.mff.cuni.cz:/MIRRORS/ftp.dsi.unimi.it/pub/security/crypt/rpub.cl.msu.edu/pc/
win/winsock/eudora14.exe

ftp.ms.mff.cuni.cz:/Security/unimi/crypt/rpub.cl.msu.edu/pc/win/winsock/
eudora14.exe

POP-Mail

ftp://ftp.demon.co.uk/pub/mail/pop/POPmail-pc/popmail/popmail-3.2.2/program/

Pegasus

http://wwwbzs.tu-graz.ac.at/software/pegasus/pegasuswebpages.html

http://www.uel.ac.uk/netw/mail/pegasus/
The best description of the best Windows free mail package

gopher://risc.ua.edu/11/network/pegasus
or: ftp://risc.ua.edu/pub/network/pegasus

ftp://ftp.shu.ac.uk/pub/netware/pegasus

'Off-line' newsreader: Free Agent

oslo-nntp.eunet.no:/pub/msdos/win3/winsock/apps/freeagent
ftp.wimsey.bc.ca:/pub/msdos/Windows/freeagent

Table 3a **continued**

On-line newsreaders:

Newswatcher

nic.switch.ch:/mirror/gopher/PCGIII/pcg3doc.zip
su1223.mathematik.uni-marburg.de:/pub/msdos/gopher/pcg3doc.zip
kth.se:/pub/tex/tools/gopher/PC_client/pcg3doc.zip
sunic.sunet.se:/pub/gopher/PCGIII/pcg3doc.zip
ftp.uab.es:/pub/msdos/gopher/pcg3doc.zip

WinVN

ftp://sunsite.doc.ic.ac.uk//packages/ibmpc/windows3/winsock/winvn926.zip
ftp.ask.uni-karlsruhe.de:/pub/infosystems/gopher/ms-windows/apps/winvn921.zip
infoserv.cc.uni-augsburg.de:/msdos/tcpip/news/winvn921.zip
rzbsdi01.uni-trier.de:/pub/pc/mirrors/tcpip/winsock/apps/winvn921.zip
ftp.istc.ca:/pub/software/pc/newsreader/winvn/winvn921.zip
micros.hensa.ac.uk:/mirrors/winsite/win3/winsock/winvn926.zip
ftp.hea.ie:/part9/cica/winsock/winvn926.zip
sierpes.cs.us.es:/pub2/dos/network/winsock/winvn926.zip
ftp.ualg.pt:/pub/computing-systems/ibmpc/windows3/winsock/winvn926.zip

WWW browsers:
Netscape

unix.hensa.ac.uk:/mirrors/netscape/navigator/gold/2.02/windows/g32e202.exe

http://the.arc.co.uk/ARC/InterNetWi/r2893842532_txt.html
http://www.toppoint.de/archiv/Netscape/Navigator-2/windows/

http://www.netscape.com

Mosaic

http://bin.gnn.com/wic/wics/browse.03.html

http://www.iwaynet.net/~andste/download.html
http://www.buss-components.com/dwlbrows.htm
http://www.ici.net/cust_pages/gillis/ftpbrow.html
http://www.bni-net.com/finney/dwlbrows.htm
http://panoptic.csustan.edu/mos.htm

http://sunsite.dsi.unimi.it/pub/www/Browsers/

Table 3b **Mac-based Internet Software**

Anarchie

unix.hensa.ac.uk:/mirrors/uunet/networking/info-service/gopher/Macintosh-
TurboGopher/helper-applications/Anarchie-140.sit.hqx

Table 3b continued

sun.dante.de:/tex-archive/tools/gopher/Macintosh-TurboGopher/helper-applications/Anarchie-140.sit.hqx

nic.switch.ch:/mirror/gopher/Macintosh-TurboGopher/helper-applications/Anarchie-140.sit.hqx

ftp.univie.ac.at:/packages/network/gopher/Macintosh-TurboGopher/helper-applications/Anarchie-140.sit.hqx

Fetch

ftp.hrz.uni-kassel.de:/pub4/mac/mac.archive.umich.edu/util/comm/fetch2.12.sit.hqx

ftp.loria.fr:/pub/mac/umich/util/comm/fetch2.12.sit.hqx.gz

Telnet

unix.hensa.ac.uk:/mirrors/uunet/networking/applic/NCSA_Telnet/Mac/Telnet2.5
sun.rediris.es:/mirror/infosystems/ncsa/Telnet/Mac/Telnet2.5
sunsite.rediris.es:/software/infosystems/ncsa/Telnet/Mac/Telnet2.5
geomat.math.uni-hamburg.de:/pub/unix/network/www/www-mosaic/Mac/Telnet/Telnet2.5
ugle.unit.no:/pub/ncsa/telnet/mac/Telnet2.5
unix.hensa.ac.uk:/mirrors/uunet/networking/applic/NCSA_Telnet/Mac/Telnet2.6
sun.rediris.es:/mirror/infosystems/ncsa/Telnet/Mac/Telnet2.6

unix.hensa.ac.uk:/mirrors/uunet/networking/applic/NCSA_Telnet/Mac/Telnet2.6/Telnet2.6.sit.hqx

Off-line mailers: Eudora

sun.rediris.es:/mirror/eudora/mac/2.0/eudora202.hqx
nic.switch.ch:/mirror/eudora/mac/2.0/eudora202.hqx
ftp.cnr.it:/pub/tp_cnuce/eudora.mirror/mac/eudora/2.0/eudora202.hqx
ftp.tuwien.ac.at:/pub/infosys/mail/eudora/mac/eudora/2.0/eudora202.hqx
vivaldi.belnet.be:/mirror/ftp.qualcomm.com/quest/mac/eudora/2.0/eudora202.hqx
infoserv.cc.uni-augsburg.de:/mac/eudora/2.0/eudora202.hqx
ftp.uni-magdeburg.de:/pub/mirror/ftp.qualcomm.com/mac/eudora/2.0/eudora202.hqx
ftp.ms.mff.cuni.cz:/Net/Mail/eudora/mac/eudora/2.0/eudora202.hqx
ftp.ms.mff.cuni.cz:/MIRRORS/ftp.qualcomm.com/quest/mac/eudora/2.0/eudora202.hqx
vcdec.cvut.cz:/pub/MIRRORS-by-URL/ftp.qualcomm.com/quest/mac/eudora/2.0/eudora202.hqx

On-line newsreader Newswatcher

unix.hensa.ac.uk:/mirrors/uunet/networking/info-service/gopher/Macintosh-TurboGopher/helper-applications/Newswatcher-20b24.sea.hqx

Telnet, or Host Presenter

Albert Finial, evolutionary psychologist, has been approached in strict confidence by the Vice-Chancellor of red-brick Teddingham university, who wants to strengthen the Psychology department there. 'But honestly, the faculty list doesn't inspire too much confidence,' he reflects to himself, as he considers his options. 'What are they really doing?'

Albert, currently examining the impact of environmental contamination on resting psyches, has already used the NISS gateway to check, unbeknownst to Professor Realitas, the comings and goings of his mentor at conferences around the world, and so it's an easy matter to look up the contemporary activities of Teddingham's Psychology faculty team. 'Oh, Fladdemus, of course, I remember that work now, he gave quite an interesting talk in Hawaii.' Albert picks up the phone. "Yes, I'll come along and look at the options, yes, fine." Just before he visits the Vice-Chancellor the following week, Albert telnets over to BIDS and checks on his own citation record, giving himself an extra boost of confidence: an additional 25 authors have cited his work in the intervening month since he last checked. Albert Finial is hot! But of course, he knew that already.

When you go out into the world of file servers, you're actually looking at the publicly available wares on that server, directly at its address. And so commands from your keyboard are eliciting responses from that server. That is to say, when you specify the directory, the server opens the one you ask it to. And then it proceeds to send the file you've asked for to your machine. So in effect, you're controlling that computer. Nice, that feeling of power from such a long way away! Telnet works, or seems to work, in much the same way.

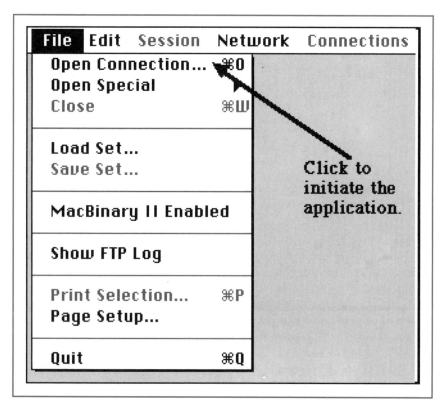

It might be easier to think of this process as a 'remote login', which is in effect what you're doing. The Telnet or Host Presenter software allows you, directly, as a person hitting the keys, as opposed to say your FTP software, to access a remote host computer, and to login as a guest, or even, if your institution has paid for appropriate privileges, as a user. Telnet software provides a wee window for you to work in, so that your communication looks a bit like Teletext; that's right, it's a simple terminal mode of operating, in simple terminal text format, and so generally felt to be less than user-friendly.

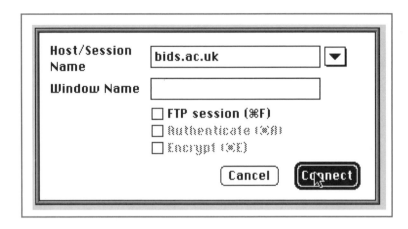

When you've logged in (for example, into the NISS (niss.ac.uk) gateway), using appropriate login name and password, which should be paid for by your institution, you can thereupon use the facilities of the remote computer to work on its own databases. Using FirstSearch software at the NISS gateway, then, allows you to search for authors (or titles or subjects) which have presented at any of the multitude of conferences worldwide in that database. A rather important tool for scientists, wouldn't you think? It's always nice to know just what old so-and-so has been up to, and how did she wangle that trip to Hawaii? Or what about BIDS? (bids.ac.uk) Always useful for checking up on your citation rating, or that of your nearest competitor.

Telnet, like FTP, is another of those very useful components of the Internet software collection which can be used as a stand-alone application, or as accessed through the console of your favourite web browser. The thing is, though, most telnet servers are now providing the same information directly to your browser in a very user-friendly manner, so you don't even have to initiate a new application — the information, or menus through which you click to get to the information, is directly available in hypertext markup language (see WWW chapters below). So there really won't be too much call, we don't think, for you to use this particular sort of software anymore. And hence the particular brevity of this section!

Unless, that is, you want access to your own institution's computers from home, for a particular purpose. For example, rather than downloading at home every bit of post that comes into my mailbox at work, I telnet directly into my university account, . . .

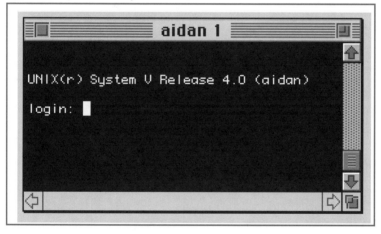

. . . access the mailing software there (it's called PINE, but that's another story), look at the list of messages, and only open the ones I want — saves a lot of phone time. In fact, this capacity to look at the whole list of messages prior to opening any of them would be a good facility to have in an off-line mailer, but sadly the freeware Eudora for Macs doesn't do this yet. So I still find the telnet software quite useful, but proably not for long.

Alternative versions of what seem to be, for practical purposes to practical scientists like me, telnet software by another name, which do some really elegant things, like Vista Xceed and eXodus (which emulate a Unix monitor on PC or Mac platforms), are beyond the remit of this particular book, but with a copy of eXodus, for example, I can do some really useful work, using the university's UNIX software, on graphing data from my own Mac at home — and they say we academic-types aren't committed workers!

We've mentioned, in an earlier chapter, the possibility of using simple telnet software to participate in Multi-User Discussion groups. Again, however, there are better applications for such participation, which don't exhibit annoying redundant messages (in Telnet, you see what you type in, and then see it again as part of the group receiving the message). And besides, as you'll have noted in that chapter, we don't reckon too many of our readers will be wanting to waste their precious time in these MUDs anyway, at least not until they can speak to, hear, and see their colleagues directly. Keyboard skills, for most people, do militate against real time chat!

The World Wide Web

Mr. Filbert, Head of Science at a progressive secondary modern, is anxiously worrying about weird stuff on the Internet. If he allows his students to expore, what strange and uncharted worlds of advertising, lust and jungle music will they find? He takes the school's modem home for the weekend to check things out for himself. Loading the easy Internet access software from the CD-ROM blanket-mailed by any of a number of service providers looking for business, he finds struggling with lots of icons, and very little science. Blunderingly, he clicks on the WWW browser icon, and after the modem buzzes and whirrs, he's suddenly presented with his service provider's home page. Oh yes, there's a search button somewhere too. Mr. Filbert tentatively keys in some words from his specialty subject, environment and ecology, as it happens, and when the search document appears, he's received a wide-ranging list of hotlinked addresses. He clicks on a couple of these addresses, and finds enough to realise that there really is a whole world of environmental information out there, notably in university syllabi. But what he really wants, is some straightforward lesson plans written by other teachers, or some up-to-date information on environmental issues around the world. How can he find these useful items?

Mr. Filbert closes his connection, and opens *this* book to *this* chapter. After he has read through the introduction, the reviews and even a bit on DIY web site construction, he has a better feeling for the medium. Eschewing our specific advice on Environmental Resources, because, in independent teacherly mode he wants to find his *own* particularly useful resources, he goes back online, confidently accesses the Altavista search engine, clicks on 'Advanced Search,' keys in 'environment' AND 'lesson plan' and incorporates 'ecology' in the 'Results Ranking Criteria'. When the Mendocino County Ecology Web link appears on the first page of search engine documents, Mr. Filbert, recognising the exhaustive interconnectedness of the Internet, decides (and why not?) to use *this* collection of friendly links, so he clicks along from 'Sustainable Earth Electronic Library' through 'Society and the Environment' to 'The Green Teacher.' Paging through the online articles

> **in this handy magazine, he finds some dynamic classroom ideas in the Renewable Energy section, particularly for Grades 10–12, the American equivalent of the British GCSE. And he's found it all by himself! Mr. Filbert creates a bookmark for his 'hotlist' and resolves to come back directly to 'The Green Teacher' on his next online session.**

This is where all this Internet stuff gets really interesting. The World-Wide Web (WWW) is a cornucopia of text, images data, sound and even video. Much of it is trivial, however an impressive amount will be of direct use to you. We'll start off by telling you what the WWW is and the sort of things you will find there. We will move on to how you can access all of this material and then we'll show you how you can add to it by creating your own pages on the WWW. We won't attempt a comprehensive listing of scientific resources on the WWW — there's just too much. What we have done instead is to list and review the best sites for each subject — sites which, in turn have links to a large number of other sites. Hold on you say "sites? links? — you're getting ahead of yourselves again — you haven't told us about these yet." Ah. Neither we have. Right, there now follows a brief description of the World-Wide Web.

Imagine that you have a document which you would like others to see. You could make it available via your university's FTP server or get it published in a journal (as you are now an Internet enthusiast you would, of course, prefer the electronic version). However both of these share the same limitation — they are two dimensional. For example, supposing you refer to other documents in the text. The reader will have to look at the list of references at the back of the document and, if interested in following it up, obtain the referenced document. There's no avoiding that with paper documents, but it is a needless limitation with electronic ones.

The WWW removes this limitation because it consists of hypertext documents. Now, of course, you want to know what hypertext means. Imagine that when reading this page all of the terms you are unfamiliar with were highlighted and that clicking on them with the mouse would take you to definitions of them. In other words you would have not one, linear document but several interlinked documents. This, in essence, is how the WWW works but on a much larger scale.

In a document on the biology of birds for example, a click on a highlighted term like 'migration' could connect your computer to another document specifically on this subject — but this time on a different computer somewhere else in the world. It is quite possible for one document on bird biology to be linked to samples of bird song on a computer in Canada, migration records from a field station in Iceland, satellite pictures in Japan and many, many other sites. It really is a world-wide web of information.

And documents themselves aren't restricted to plain text. It is possible to produce illustrated documents as impressive as anything available on paper — and much more cheaply, since the only resource you use up is disk space on your computer. This makes the WWW a very powerful tool for communication. Previously the Internet was a different way of text-based communication. The WWW shows its ability to be something more, a new 3-dimensional way of communication combining text, graphics, sound, video and even software programmes. If that sounds a bit intimidating, don't worry — most sites you will come across will still be in the familiar text and picture format, with the added feature of links to other documents elsewhere. However, it is important that you realise where the WWW is headed — there are exciting times ahead.

There is a drawback to all this of course. Sending pictures and sound across the web means sending large files through a system designed for sending text-based data. That means that the net slows down considerably at certain times of the day due to the sheer weight of data. Try visiting a site in the morning and then in the afternoon once the USA has started work and see the difference for yourself. The slowdown is such that if your Internet access is via a modem, the modem speed will cease to be a bottleneck — instead it will be the transatlantic cables.

So what do you have to do to get there and what will you actually see? One of the nice things about the WWW is that it was designed from the beginning to be user friendly. Email, Usenet etc. all started off with horrible UNIX software and then had nice, friendly Windows and Mac software developed for those without the time or inclination to take a degree in computing — and sometimes it shows. The WWW has only ever been accessible using browser software.

There are many different types of browser software available but we reckon that there's only 2 or 3 that you need to bother about. The first one is the original one — Mosaic. This programme is developed by the supercomputing laboratory at the University of Illinois. You can obtain the latest version by FTP as noted in the FTP Chapter. And here's the really good news — it's free. That's right, free — as in you get a wonderful piece of software for nothing. The second browser you need to know about is Netscape Navigator. This is the best browser currently available. But it isn't free, well not for everybody anyway but we'll come back to that.

217

Netscape was developed by Marc Andreesen who did some of the original work on Mosaic. Netscape Navigator is consistently at the leading edge of what is possible on the WWW and therefore is the browser used by something like 80% of WWW users. It is available by FTP from the address mentioned in the FTP chapter. So do you have to pay to get it? No, not at the time of writing. Anyone can download the latest version of Netscape from the FTP site. However you have a moral (and legal) obligation to pay for it unless you are in one of the categories of people who the nice people at Netscape allow to use it for free. At the time of writing this included academics, students and charities. The third browser we'll mention is one that we've never used — Microsoft's Internet Explorer. Microsoft being what it is, Internet Explorer is being heavily promoted and has its fans. We're not going to tell you where you can get it though; we think Bill Gates has more than enough money as it is.

All of these browsers may look slightly different but they all work in exactly the same way. They connect your computer to other computers using the same TCP/IP protocols we mentioned in the introduction. However all of that happens beneath the bonnet. To begin with, you'll find your browser, after you've downloaded it from an FTP site, or procured it as part of some bundle, in its own special folder:

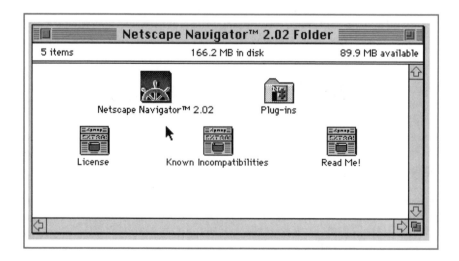

Click on the Netscape (or other browser) icon, and what you see is something like this:

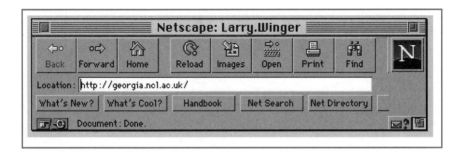

Of course, it will have a different title, and no doubt will be opening up a file at someplace like: http://www.netscape.com/ but that's a different story. What you usually see is the button menu bar, to begin with, so you can click the 'Open' button, or you can use the standard menu at the top of the application:

Whatever, the point is that you need to point the browser or direct it to a specified address in the World Wide Web:

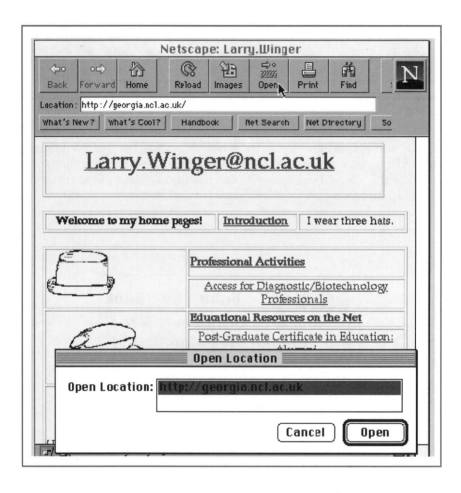

You type in the address of a website of interest — say http://
www.georgia.ncl.ac.uk (you'll notice that all WWW addresses have the
same 'http://' prefix). Then you'll see something like the above. In fact
what you're looking at is Larry Winger's home page. What this contains
is information about Larry (go on have a look, he's an interesting guy),

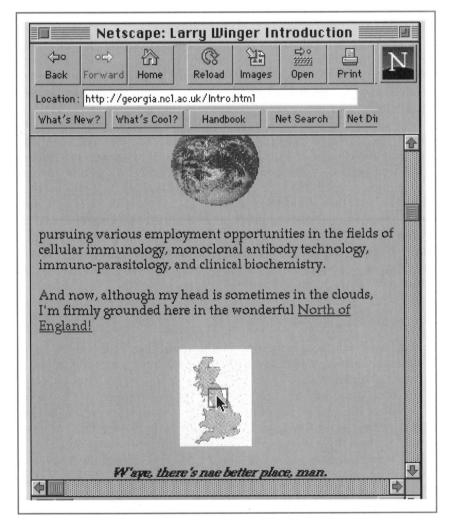

in addition to other documents he has produced — a on-line version of
a recent publication perhaps — and links to other computers all over the
world with resources relevant to his interests. Every time he clicks on his

browser icon the first thing that comes up on the screen are these links — because they are the ones he uses regularly. One of the sites that he uses regularly is the Errata site for this book; it may not necessarily be linked through his home pages, but rather it may be a 'narrowcast' publication, with the address given out only to those people who read this book:

http://www.compulink.co.uk/~embra/errata.html

It really is as easy as that — you just point and click. That in itself would be great but there's more. As we've said elsewhere, WWW browsers combine many of the features of other Internet software. So Netscape, for example, can be used to send email, read and post to Usenet newsgroups and retrieve FTP files.

So you've a vague idea now what this WWW thing is and what browsers are. You have Netscape or Mosaic on your PC and you want to explore. You click the icon and the programme fires up. Now what? Now you go to the next section of this book: the subject-oriented website reviews. If your particular specialist subject isn't covered then we'll show you how to find information about the subject of your choice. In an hour's time you'll be switching back and forth around the globe like a veteran.

Now if this book doesn't have a coffee ring on it somewhere within a week, we have obviously failed somewhere along the line. But to prevent too much wear and tear, we've linked all of the WWW sites reviewed in this chapter to a special website at

http://www.compulink.co.uk/~embra/ifswww.html

This means that although the sites might move, there'll always be an up to date copy available to you on-line — and if there's a dead link, we'd certainly like to be told.

"Hold on" you may be saying by now "all this sounds a bit too good to be true to me. "How did all the wonderful information you're talking about get there in the first place? Can it really be free? And if it's free does that mean that it isn't very good?" All good questions — we would expect no less from a scientist. Many web sites are the work of one individual who is interested enough in their subject to put together a site. They may want to display some of their own work for example and having done so have decided to pull together other sites which deal with the subject. They may do this for themselves or on behalf of their department or organisation. Individual enthusiasts are responsible for providing much of the information available on the web for scientists and where possible we have credited them in the reviews. Increasingly though, web sites are professionally designed. A company might want to provide a site which provides a useful service and at the same time allows it to advertise itself. It's the same principle as sponsoring a sports event except that they can target a very specific audience. Individual enthusiasts are still the lifeblood of the net though.

Once you've visited a few sites, you too won't be content with just being a spectator. You'll want to establish your own website as a shop window to the world for your work and that of your laboratory. After the reviews we will show you exactly how to build your own basic web page. You'll be amazed how easy it is. You'll also be amazed at how many complete strangers from all over the world will come and look at it.

SEARCH ENGINES

However, before we get to the reviews, we really have to mention what is perhaps *the* most important component of the WWW, and that is the amazing 'search engines'. It's the search engines that allow you to find that critical piece of information from all the dross, and without them the WWW just wouldn't be the same. There are two kinds of search engines, and you use them in different ways.

The first type of search engine works a bit like a library. WWW sites are arranged into headings and sub-headings and you can browse to see if there's anything in the collection that is of interest to you. We're all familiar with directories, subdirectories, and sub-sub-directories (or, as we Mac-users like to say, folders within folders). The best and most popular of its ilk is the Yahoo site. Here at:

http://www.yahoo.com/

you can click on and page through the directories and sub-directories until you find a list of WWW sites for the subject you are interested in . . . much like these ones:

YAHOO Directories:

Arts and Humanities
Architecture, Photography, Literature . . .

Business and Economy
Companies, Investments, Classifieds . . .

Computers and Internet
Internet, WWW, Software, Multimedia . . .

Education
Universities, K-12, College Entrance . . .

Entertainment
Cool Links, Movies, Music, Humor . . .

Government
96 Elections, Politics [Xtra!], Law . . .

Health
Medicine, Drugs, Diseases, Fitness . . .

News and Media
Current Events, Magazines, TV, Newspapers . . .

Recreation and Sports
Sports, Games, Travel, Autos, Outdoors . . .

Reference
Libraries, Dictionaries, Phone Numbers . . .

Regional
Countries, Regions, U.S. States . . .

Science
CS, Biology, Astronomy, Engineering . . .

Social Science
Anthropology, Sociology, Economics . . .

Society and Culture
People, Environment, Religion . . .

Similar directories, with more local flavour, are found in the various National Yahoos: Canada — France — Germany — Japan — U.K. and Ireland

(http://www.yahoo.co.uk).

Now, this system has its advantages — not least that it allows you one of the great pleasures of library and bookshop browsing — finding other useful things that you didn't even know you were looking for. However, it shares the same limitation as libraries — it is put togther by humans. Yahoo is dependent on recommendations from its employees and users for additions and corrections, so there will always be very good sites that it misses — a fair proportion of our reviewed WWW sites are not listed on Yahoo for example.

Here's an example of how you could use a directory of WWW sites like Yahoo. You *can* find Newcastle University by clicking through the Yahoo Directory, moving into say the UK and Ireland mirror site, then into the Universities link under the Education menu, after you've specified the United Kingdom directory, and after waiting a while for the list of all the universities in UK to crank along, you'll finally find the University of Newcastle upon Tyne, or Newcastle University as is.

- Loughborough University@
- Luton University@
- Manchester Metropolitan University@
- Manchester University@
- Middlesex University@
- Napier University@
- Newcastle University@
- Nottingham Trent University@
- Nottingham University@
- Open University, The *(5)-* providing open courses.
- Oxford Brookes University
- Oxford University@
- Paisley University
- Plymouth University@
- Portsmouth University@
- Queen's University@

Interestingly, the search terms in Yahoo don't seem to work very well — plug in 'newcastle' on the entry page, and you get only 7 references back, none of which are what you want. You won't receive anything for the compound key word entry: newcastle clinical biochemistry. Well! Once you've entered Newcastle University www space, though, moving hierarchically through all those directories, you can then move through another set of directories, from faculties, to medical school, to Clinical Biochemistry, and finally you might find one of us! So you've got here in the end!

The UK Index is arranged similarly, and works on a similar directory principle.

http://www.ukindex.co.uk/uksearch.html

Arts	Ireland	Recreation
Business	Jobs	Reference
Computing	Language	Regional
Culture	Law	Religion
Education	TV, Radio, Video, Films	Science
England	Magazines, Newspapers etc	Scotland
Environment	Music	Sports
Health	Nature	Travel
History, Genealogy	News	Wales
Index	Politics, Government	Weird
Internet	Professional	

Some of us find this directory approach quite tiresome, but it has the advantage of actually getting you to a site in a logical fashion. All too often, however, in these search facilities which have been set up and are maintained by humans, you just can't immediately find what you're looking for, especially if you use keywords and you think you should be able to get there right now! 1, 2 mouse clicks at the most! And that's frustrating!

The second type of search engine takes a different approach. No directories and sub-directories, these engines are run by robots. There are currently several really useful robotic search engines, and no doubt you can find more, but these are the ones we really like and use.

The Altavista Search Engine is at http://www.altavista.digital.com/

The Lycos Search Engine is at: http://www.lycos.com/ or as originally: http://lycos.cs.cmu.edu/

The WebCrawler Search Engine is at: http://webcrawler.com

The Magellan Search Engine is at: http://www.mckinley.com/

A fifth search engine is also worth mentioning: DejaNews at: http://www.dejanews.com/ DejaNews doesn't search the WWW, it searches Usenet posts. Incidentally, you can also search contemporary Usenet postings at some of the other sites too. But hey, you can never have too many search engines, when you've gotta, just gotta find that crucial item.

So supposing that, in one of these robotic search engines, which have been built up both by clever robotic programmes searching out new and interesting sites, as well as by direct submission of site addresses by individual people (more on this below), I key in my three search terms "newcastle clinical biochemistry. In only moments I'm presented with 10 citations, of which the first is Newcastle University in Australia, the next is the Association of Clinical Biochemists in Birmingham, but the third is me, since I maintain the Clinical Biochemistry department's pages. So I've found meself at last. Sometimes robots really are better than people. Using the appropriate key words, and narrowing your search, sometimes using alternative search engines, if the one you're using is a bit idiosyncratic in its understanding, will quite often get you the information you desire in a very short space of time. The keywords can also be entered with appropriate Boolean logic (that's as in X AND Y, or X NOT Y).

This symbiosis between humans and robots can be quite frightening!

The other good thing, however, about the robotic search engines is that you can also use them to search current postings on the Usenet Newsgroups. So, web or newsgroups, we'd have to say that the robotic search engines are just great. We've already mentioned, in the FTP Chapter, use of different search engines for FTP sites, and files found therein. Some of them, like Altavista, already have mirror sites outside their North American home base. They're only going to get better and better!

ACADEMIC JOURNALS ONLINE

'Classic' or Conventional Scientific Journals On-Line

An increasing number of scientific journals, with full articles as part of the service, are being presented on the WWW. In most cases, your institution must be part of an agreement with the publishers, in the UK referred to as the CHEST agreement, and your department will have an identification code as well as a password, which you will need in order to access the full texts.

Here at the University of Newcastle, our excellent library facilities provide a great deal of useful information on accessing these journals, which can be found at the URL:

http://www.ncl.ac.uk/library/

On the other hand, access to contents pages, and in many cases, to abstracts, is available for free from the Internet, to any interested browser.

Most of these services also offer their internal search engine, in case you want to look for a specific author or title. For example:

Academic Press: http://www.janet.idealibrary.com/
or: http://www.europe.idealibrary.com/

Chapman & Hall: http://hermes.chaphall.co.uk/
Elsevier Press: http://www.elsevier.nl
IEEE: http://www.ieee.org/
Institute of Physics Publishing: http://www.iop.org/
Springer Journals: http://www.springer.de/server/svjps.html

In addition, of course, we've already mentioned BIDS (http://www.bids.ac.uk) in the telnet section, which is also linked through the NISS information gateway (http://www.niss.ac.uk), as is the Firstsearch service. For really useful searches, using these services, of course, your institution must have an account, your department an identity and a password. Blackwells Publishers also offer their publications online, through the BIDS service (with the same ID and password), at:

http://www.journalsonline.bids.ac.uk/Journals Online

Of course, we'd be quite remiss if we didn't note the home page of our publishers, Harwood Academic a member of The Gordon and Breach Publishing Group, and their online information on some 300 journals in the physical, medical and life sciences spheres:

http://www.gbhap.com/

Most of the journals that are entirely online use the Acrobat publishing format, so that you download a '.pdf' file, and then read it (ie the full paper, with figures, tables, references) with Acrobat reader software. The reader software is free, and is always available for download from any site which offers .pdf files. We've already dealt in some detail with finding useful software via FTP servers in the FTP Chapter, and this is an occasion when you can really use the advantages of ftp, to get that hot paper onto your computer immediately! Your www browser can be configured to find the Acrobat reader application when it has downloaded a .pdf file, so that reading the paper should be as easy as point and click . . . oh, and wait . . . . and wait some more?

We've said it before, we'll say it again . . . when the net is slow (as when North America wakes up) try sites closer to home, and you'll be glad you did!

Electronic Journals

Increasingly, you'll find that journals available *only* in electronic format are popping up around the net. We don't think it useful (in *this* edition of our book) to catalogue all these journals, since you will undoubtedly be able to find them through the subject-specific links itemised and reviewed below. Traditionally, of course, scientists like to see their work in hard copy format, bound in a tree-hungry volume. But what, we're forced to ask, is the point, really? Well, journals are kind of handy to hold in your hand, sure, but if you're in front of the computer monitor anyway, it's a lot easier to find what you want in a couple of clicks, then to hie off to the library, search through the stacks, retrieve that one volume that has wandered off to the photocopier staging area, wait in the photocopying queue, and then wreck your eyes deciphering the smudges.

The advantages of WWW journals are potentially huge. The costs of production are low — so low that many of them are provided free by societies as a service to their members and to the scientific community in general. The turn-around time at which received papers can be edited and published is much lower than for paper journals and yet the all-important peer-review process is still there as a guarantee of quality. So what's the problem? At the moment, people are still circling warily around the WWW journals — they're waiting to see what others do. It's fair to say that the traditional paper-based journals still hold more prestige and, obviously, people want their publications to have as big an impact as possible. However, we predict that that will change quite rapidly. All it takes is for a few key papers to appear in electronic form and the mould will be broken forever. The advantages in speed, visibility and increased feed-back will lead to WWW journals competing on an equal footing with non-WWW ones. We predict that this, allied with the library budget cuts that seem to be endemic at the moment, will mean that quite a few of the smaller journals will simply not survive in printed form.

We say: on with the electronic revolution, keep up the quality, the peer review, and the brilliant as well as the pedestrian science, and let's get on with it!

Reviews of WWW Science Sites

Hopefully your appetite is at least partially whetted. All this WWW sounds quite exciting and you've picked up enough jargon to bluff your way through a dinner party or coffee break. But the key question is: is it actually useful? In these reviews, we hope to show you the sort of science resources that are available to you via the WWW. Prepare to be amazed; or at least mildly surprised.

ABOUT OUR REVIEW SYSTEM

We haven't even tried to review all of the science sites on the WWW. We'd need something the size of a telephone directory and a great deal more free time than we have. What we've done instead is to take representative sites from a range of subjects. These sites have been chosen because we think that they are well designed, have a good amount of useful information, will connect you to other useful information on the same subject and are reasonably permanent.

Agriculture

Ag-links: http://www.gennis.com/ag-links.html

A long list of agriculture related WWW sites compiled by the Gennis Agency — presumably as a shop window for their agri-business PR services.

Not Just Cows: http://www.snymor.edu/~drewwe/njc

Nicely laid out guide to agriculture resources on the Internet. This site is a labour of love by Wilfred Drew. Has comprehensive information on mailing lists, electronic journals, WWW sites and more.

World Agricultural Information Center:
http://www.fao.org/WAICENT/waicent.htm

This site belongs to the Food and Agriculture Organisation of the United nations (FAO). Its purpose is to make the FAO's library of information available to whoever wants it, from governments to individuals. Statistical and textual information relating to agriculture, fisheries, forestry, nutrition and rural development can be accessed here. It's difficult to imagine how this could have been done pre-Internet.

Anatomy

American Association of Anatomists: http://faseb.org/anatomy/

Surprisingly perhaps, anatomists are at the forefront of scientists exploiting the possibilities the Internet offers. Several sites have interactive resources which look at whole bodies or specific organs in some detail. You can access them all from this site.

The Visible Human Project:
http://www.nlm.nih.gov/research/visible/visible_human.html

The project is the creation of complete, anatomically detailed, three dimensional representations of the male and female body. Impressive progress has been made on this mammoth project, as you will find when you visit. How could all this information have been made available without the Internet?

Anthropology

Anthropology Department, Laurence University:
http://www.lawrence.edu/dept/anthropology/

Information about the Department, links to Anthropology resources but most importantly this site is home to Classics of Out(land)ish Anthropology, featuring reviews of sites which abuse the good name of anthropology — lost civilisation weirdness this way.

Anthropology Resources on the Internet:
http://www.nitehawk.com/alleycat/anth-faq.html

Aside from one picture of a skull this site is essentially one long document (34 A4 pages) listing a huge amount of Anthropology-related information. This includes WWW sites but also email lists and electronic journals. It's a bit too unwieldy to use as a regular site — the best thing is to keep it on your hotlist but prune out particular sites of interest for your own use.

Archaeology

Archaeology Resources on the Internet:
http://www.arts.gla.ac.uk/www/ctich/archlinks.html

An excellent set of annotated archaeology links: lecture courses, virtual tours of sites, and major resource collections.

Astronomy, Astrophysics and Space Science

Adventures in Astronomy: http://mindspring.com/~thendrix/

Complete with a nice picture of Orion's Horsehead nebula, this site is designed for both amateur and professional astronomers. Links to Hubble images, a guide to the planets and much more.

Ames Research Center, Space Science Division:
http://www-space.arc.nasa.gov/division/

Starts off with an interesting illustration bearing the caption: "This illustration represents the host of natural phenomena which collectively have created life as we know it." We were quite cheered by this, because we didn't think that you were allowed to say things like that in the USA anymore. Lots of information on the Division's R&D "dedicated to understanding of the origin and evolution of starts, planets and life". Good links too.

AstroWeb: http://www.stsci.edu/net-resources.html

A searchable WWW site listing Astronomy resources not just on the web but all over the Internet. The annotated WWW resources are the star of the show though (sorry about the pun). The site is easy to access because it is virtually graphic free, allowing you to straight to the information you want. In fact the only graphic is a reproduction of Van Gogh's 'Starry Night' — a nice touch we thought.

The High-Energy Astrophysics Learning Centre:
http://heasarc.gsfc.nasa.gov./docs/learning_center/

This well-designed site has more information about high-energy astrophysics than most of us can imagine. Contains basic guides, satellite data, software and much more.

The Planetary Rings Node: http://ringside.arc.nasa.gov/

This site archives and distributes data relevant to planetary ring systems. This is divided into fun stuff (images and animations of planetary ring systems) and serious stuff (Voyager information and data and how to use it). Just think how difficult it would have been to make such information generally available pre-Internet.

Radio Pulsar Resources: http://pulsar.princeton.edu/rpr.shtml

If pulsars are your subject then look no further than this page. Image-free links to pulsar publications and pre-prints, pulsar scientists and laboratories world-wide. All pulsar life is here.

Theoretical Astrophysics: http://wonka.physics.ncsu.edu/

This site is based at the Department of Physics at North Carolina State University. Contains good general astronomy links as well as information about the Departments own research interests. Includes a link to NASA's searchable index of astrophysics research papers.

Biochemistry

Biochemistry On-Line: http://www.arach-net.com/~jlyon/biochem/

Subtitled "Biochemistry — the building blocks of life", this site contains biochemistry documents, on-line courses, impressive molecular modelling and much more.

PUMA:
http://www.mcs.anl.gov/home/compbio/PUMA/Production/puma_graphics.html

PUMA is an acronym for Phylogeny Metabolism Alignments. This wonderful site is a good example of what the WWW has made possible. It is a database of metabolic pathways, with links to genomic and other data about the enzymes involve. You can also find information about substrates and make comparisons between different organisms. In other words it integrates information from a number of different sources into one valuable resource. This really is very impressive.

Bioinformatics

Biologist's Control Panel: http://gc.bcm.tmc.edu:8088/bio/bio_home.html

The Biologist's Control Panel has links to a large number of nucleic acid and protein databases and their search engines. From here you can search for specific sequences and compare them with others. Links to other bioinformatics sites.

Biophysics

Biophysical Society: http://molbio.cbs.umn.edu/biophys/biophys.html

Information about the Biophysical Society and biophysics plus links to other biophysics sites on the WWW.

Biology

BIOSCI/Bionet: http://www.bio.net/

Has a few links to other biology sites but the important thing about this site is that it contains a searchable archive of every post ever made to a bionet newsgroup. You can also use it to read current newsgroup posts and to send your own messages. Has links to WWW sites of companies which sponsor BIOSCI — you are encouraged to follow these links and congratulate them for their support. Without that support this invaluable resource for biologists could not continue.

Project BIO: http://biotech.zool.iastate.edu/Project_BIO/Homepage.html

A biology WWW site for students and educators, dedicated to develop and share biology resources via the Internet. Includes on-line courses and seminars in biology. These are in multimedia format with text, images and audio.

Biotechnology

Biotech BiblioNet: http://schmidel.com.biotech.htm

Subtitled "a monthly bibliography and resource centre" this site is a boon for anyone trying to keep up with the latest developments. Biotech citations from a range of journals are listed, often with links to abstracts and to the email addresses of the authors. And it's searchable too. Also features a good set of links to other sites and (nice touch) a list of sites which feature a link to this one.

Biotechnology Dictionary:
http://biotech.chem.indiana.edu/pages/dictionary.html

This is an illustrated on-line dictionary of biotechnology terms. Just type in the term you want to search for.

Biotechnology Information Centre: http://www.nal.usda.gov/bic/

This site is an information centre for agricultural biotechnology run by the National Agriculture Library of the US Department of Agriculture. It has links to a definitive collection of biotechnology papers, newsletters, patents and education resources, as well as other agricultural biotechnology sites world-wide. Also features resources on specific subjects like bovine somatotrophin and performance standards for GMO research.

Chemistry

The Analytical Chemistry Springboard:
http://www.anachem.umu.se/jumpstation.htm

From Umea University in Sweden, this site offers a very good selection of links to analytical chemistry resources, from atomic spectroscopy to x-ray spectroscopy. Also contains to suppliers of analytical chemistry reagents and software, on-line tutorials and journals.

Chemical Engineering:
http://www.che.ufl.edu/WWW-CHE/index.html

The WWW Virtual Library section on chemical and process engineering — links to resources in a variety of categories from ceramics to water technology, and to organisations and laboratories all over the world.

Chemistry on the WWW:
http://badger.ac.brocku.ca/~gt95ab/chem/chem1.html

A compact collection of chemistry WWW sites, each annotated and rated. The system of rating is 1 to 10 amu (atomic mass units) for the quality of information and a series of metals are used to rate the quality of the website itself. We need more sites like this.

Chemistry Resources:
http://www.nie.ac.sg:8000/~wwwchem/l-chres.htmlx

This is a well designed collection of chemistry resources on the Internet, based at Nanyang Technological University in Singapore. Includes chemistry software, mailing lists, on-line journals and much more.

Chemistry Teaching Resources:
http://www.anachem.umu.se/eks/pointers.htm

A comprehensive collection of links to chemistry teaching resources on the Internet: demonstrations and experiments, graphics, software, safety information and much more.

Chemistry Tutor: http://tgd.advanced.org/2923/html/index2.html

Another good collection of chemistry educational links.

Chemist's Art Gallery: http://www.csc.fi/lul/chem/graphics.html

A spectacular site full of chemistry-related visualisations and animations and links to more of the same elsewhere. Everything from animations of small molecule diffusion in polymers to vibrational modes in benzene. Absolutely fascinating — and very useful if you are a chemist.

Chemweb: http://www.kern.com/~chem-man/chemweb.htm

A WWW site devoted to chemistry. Well designed and not too heavy on graphics. Features include an interactive periodic table and molecular models for a range of compounds. Has extensive links to other chemistry pages around the world.

Electrochemical Science and Technology Information Resource (ESTIR):
http://www.cmt.anl.gov/estir/info.htm

A comprehensive collection of electrochemistry resources available on the Internet. The collection was mainly developed through the FAQ for sci.chem.electrochem

Internet Chemistry Resources:
http://www.rpi.edu/dept/chem/cheminfo/chemres.html

This is another collection of chemistry links. Rather than taking the comprehensive approach it takes a more selective approach. It is also searchable.

MAG-NET Home Page:
http://www.chemistry.uakron.edu:8080/cdept.docs/nmrsites.html

A nicely designed collection of magnetic resonance resources on the Internet.

This site is a good example of how it is possible to create web pages that look good without being overburdened with graphics.

Material Safety Data Sheet Searches:
http://research.nwfsc.noaa.gov/msds.html

An invaluable resource for those responsible for lab safety documentation. This is a searchable index of MSDS information. Not complete but a useful first place to look.

MS Links: http://www.sisweb.com/mslinks.htm

Information and links relevant to mass spectroscopy, compiled by SIS. Includes on-line journals and tools such as the Exact Mass calculator — enter a chemical formula and it will calculate the exact mass and isotopic abundance for the compound. Many other useful features.

Organic Chemistry Resources Worldwide:
http://heme.gsu.edu/post_docs/koen/worgche.html

A large collection of links relevant to organic chemistry. Includes the sort of links you might expect — information on compound properties, purification and nomenclature — and other useful things too, like information on patents and article writing.

Polymers DotCom Home Page:
http://www.polymers.com/dotcom/home.html

Despite the confusing (but no doubt amusing when they thought it up in the pub) address this is a very impressive site. It is sponsored by a number of companies from the plastics industry, which no doubt contributes to its professional look — or perhaps resulted from it, who knows. In any case, what you will find is a compendium of information and links on polymer science. There are links to polymer resources elsewhere on the Internet, a basic guide to plastics and much more. There is even a children's section showing you how you can impress your kids by making that slime stuff.

Sheffield ChemDex:
http://www.shef.ac.uk/uni/academic/A-C/chem/chemistry-www-sites.html

A comprehensive listing of nearly 2,000 chemistry resources on the Internet.

WebChemistry: http://www.latrobe.edu.au/www/wechem/

This definitive-looking collection of chemistry links is divided into academic sites, commercial sites, software, resources and jobs. It also has mirror sites in the UK and USA, where there are people maintaining the site. The best collection of general chemistry links.

WebElements: http://www.shef.ac.uk:80/~chem/web-elements/

This is a hypertext periodic table — click on an element and you are connected to information on it in a variety of categories from nuclear to biological. The table is constructed by Mark Winter and made its first appearance in 1993, making it the WWW equivalent of a Giant Redwood.

Earth Sciences

Earthrise: http://earthrise.sdsc.edu

A collection of images of the Earth's surface as viewed from the space shuttle. Can you find your house?

Earth Science Institutions Directory: http://scilib.ucsd.edu/sio/inst/

Based at the Scripps Institution of Oceanography library, this site comes with the reasonable enough disclaimer that it is selective for SIO's information and needs. To which we can only respond that its needs must be great, because this is a very big list of earth science sites, with links to datasets, research, images and all the good things that we expect from the net.

Earth Science Resources:
http://www.geosci.unc.edu/web/ESresources/ES12795.html

An excellent collection of earth science links. What normal person could resist visiting the 'Seismogram of the Day' page? If it has a fault it is that it is a bit heavy on the graphics.

Earth System Science Resource Center:
http://www.thompson.com/rcenters/earthnet/earth_sci.html

Sponsored by the Wadsworth Publishing Company this is an impressive site with electronic field trips and links to resources on the lithosphere, hydrosphere, oceans, atmosphere, meteorology and much more. One for the hotlist.

GeoSim: http://geosim.cs.vt.edu/index.html

GeoSim is a joint project between the Geography and Computing departments at Virginia Tech. It consists of an on-line geography course with tutorials and simulation programmes based on the tutorial concepts.

GeoWeb: http://wings.buffalo.edu/geoweb/

This aims to be the home page for those interested in Geographic Information Retrieval. It includes and information and links concerning GIR theory, research and data. Links to a huge amount of mapping data.

NSSDC Earth Science Data:
http://nssdc.gsfc.nasa.gov/earth/earth_home.html

The National Space Science Data Centre has 3.4 terabytes of earth science data, much of it available here to users.

On-line Resources for Earth Scientists (ORES):
http://www.calweb.com/~tcsmith/ores/

Another geoscience, meteorology, geology and geography listing. Also very good. Decide which one suits you best.

The Virtual Geomorphology: http://hum.amu.edu.pl/~sgp/gw/gwl.htm

A comprehensive collection of geomorphology resources in English and Polish.

Environmental Science

Biodiversity and Biological Collections WWW Server:
http://www.keil.ukans.edu/

A searchable index of taxonomic and systematics data, collections departments and so on. A good collection of information on a site that is attractively put together without overdoing the graphics.

EOS-RAM Home Page: http://boto.ocean.washington.edu/eos/eos.html

This is the site for a project arising from NASA's Earth Observing System, researching into the biogeochemistry, hydrology and sedimentation of the Amazon river basin. A great deal of useful data for anyone interested in this area.

Enviro$ense: http://es.inel.gov/

Enviro$ense — yes they really do spell it, for reasons best known to themselves, with a dollar sign instead on an 's' — is funded by the Environmental Protection Agency to provide information on technical and regulatory issues related to pollution prevention and environmental compliance. This site draws in a huge amount of data from different resources and makes it available to any scientists interested in these issues.

GAP Analysis Home Page: http://129.101.133.222/gap/

Overview of and data from the Gap Analysis Programme which seeks to provide regional assessments of the status of vertebrate species and land cover types, and apply it to land management.

Global Change Master Directory: http://gcmd.gsfc.nasa.gov/

This site is a searchable index of earth science, environmental, biosphere, climate and global change data available to the international scientific community. This really is a very impressive site. It's difficult to see how all of this data could have been gathered together and made available so easily to many people without the Internet.

Population Ecology:
http://gypsymoth.ento.vt.edu/~sharov/popechome.welcome.html

A listing of sites with information relevant to quantitative population studies. Has a great many links to models, data for modelling, on-line lecture notes and much more of interest to those in this field.

US LTER: http://lternet.edu/about/program/what.htm

The Long Term Ecological Research program is a collection of US-based projects on long-term ecological phenomena. History and results are available here along with links to similar international projects.

Engineering

CEnet: http://www.cenet.org/

CEnet is the Civil Engineering Research Foundation home page and has a well organised collection of research related civil engineering links. It also has a 'members only' section to contents of which we can only guess at. Judging by engineers of our acquaintance it is probably full of 'glamour' calendars.

EELS: http://www.ub2.lu.se/eel/

EELS is the Electronic Engineering Library, Sweden. This is a collection of engineering resources on the Internet, each of which is quality assessed. It was still under construction the last time we visited. By the time you read this it will be fully on stream. It uses much the same classification as EEVL (see below) but being Swedish also has a section on Polar Research and Cold Region Technology.

EEVL: http://eevl.icbl.hw.ac.uk/

EEVL is Edinburgh Engineering Virtual Library, based at the city's Heriot Watt University. This is a collection of general engineering links which is very well designed. The different links are categorised into Chemical, Civil, Design, Electrical, General, Environmental, Materials, Mechanical and Offshore engineering. It also contains an archive of engineering Usenet newsgroups. The almost graphic-free interface allows for fast access and, best of all, the links are annotated, telling you what resources you can expect to find. The Egyptian Eye of Horus logo had us puzzled until the penny dropped. However, we can forgive them their atrocious pun because this is a model example of how such sites should be.

Electronic Engineers Toolbox: www.eetoolbox.com/ebox.htm

A collection of frightening looking resources for electrical and design engineers. If digital signal processing and embedded systems are your area then you will find many links to relevant resources here.

Nuclear Engineering: http://neutrino.nuc.berkeley.edu/NEadm.html

This site is the Nuclear engineering part of the WWW virtual Library. It is a definitive collection of nuclear engineering resources including research pre-prints, a bulletin board, and many links to laboratories, organisations and data.

Petroleum Pages:
http://www.pe.utexas.edu/Dept/Reading/petroleum.html

This site is concerned with petroleum and geosystems engineering. It has a nicely informal touch and, as well as links to all of the resources you might expect (research, software etc.), it also contains information about scholarships for school students interested in the subject.

Entomology

Entomology Index of Internet Resources:
http://www.public.iastate.edu/~entomology/ResourceList.html

At the time of writing this page had links to over 700 entomology resources on the Internet. Web sites, mailing lists, job opportunities — this is a real one-stop shop for the busy bees of the Entomology world.

Forensic Entomology Home Page:
http://www.uio.no/~mostarke/forens_ent/forensic_entomology.html

Slightly off-beam perhaps, this site is effectively an on-line textbook on forensic uses of entomology. All the information you might conceivably want on the use of insects to solve murders and catch smugglers plus links to other forensic entomology pages (yes, surprisingly there is more than one — you learn something new everyday eh?).

Medical Entomology and Insect Physiology Index Page:
http://www.pe.net/~chamcham/medent/

This is Chuong Huynh's collection of information and links for medical entomology. You will find everything from images of mosquitoes to Plasmodium DNA sequences.

Evolution

Enter Evolution: http://www.ucmp.berkeley.edu/history/evolution.html

This first rate site collects writings of Darwin and other scientists from Aristotle onwards which are relevant to evolution and natural selection.

It functions as a very good history of natural science text. It also has topics on systematics, taxonomy and fossils and their relevance to evolutionary theory. Should be on everybody's hotlist.

Origins of Humankind: http://www.dealsonline.com/origins/

A collection of human evolution information, links and bulletin boards.

The Origin of Life: http://users.aol.com/chinlin3/home.htm

A comprehensive set of information and links on the astronomical, chemical and biological aspects of the origin of life. The information is impressively set out — you can take a 'time machine' to different ages or look through the site in a pre-arranged sequence.

Phylogenetics Resources:
http://www.ucmp.berkeley.edu/subway/phylogen.html

A collection of phylogenetics publications, databases and software.

Genetics

On-line Mendelian Inheritance in Man:
http://www3.ncbi.nlm.nih.gov/Omim/

This is a searchable database of human genetic disorders. Each entry has detailed textual information and references, links to Medline articles, DNA sequence data and images. This really is an impressive tool for anyone in the field of human genetics or inherited disease.

Marine Biology

The Cephalopod Page: http://is.dal.ca/~ceph/wood.html

A stunning archive of textual and visual information about Cephalopods (squids, cuttlefish, nautilus and octopus) and other molluscs. Links to similar sites and Internet resources. Well worth a visit.

Marine Biological Laboratory: http://www/mbl.edu/

This is the home page of the Woods Hole marine biology lab. Amongst its many useful features is a marine specimens database incorporating images, taxonomic and genomic information. There is also much other information relating to the work of the laboratory and links to a number of marine biology sites elsewhere — some of which are pretty spectacular.

Mathematics

e-Math Home Page: http://www.ams.org/

The home page of the American Mathematical Society. This is an excellent site full of mathematics news, conferences and other professional information. It also has Mathematics on the Web, a comprehensive collection of maths links from Actuary to wavelets. The best maths index on the net.

Level Set Methods: http://math.berkely.edu/~sethian/level_set.html

Level Set Methods are numerical techniques which can follow the evolution of interfaces. It is included here because it is well designed, has basic information about the subject and has a collection of relevant movies and Java applets. In short, it's a good example of what a specific site like this should be like. Maintained by J .R Sethian at Berkeley University.

Math Database: http://www.emis.de/cgi-bin/MATH

A searchable database of mathematics abstracts from 1931 onwards. Unless you want to pay, you're limited to 3 answers.

Mathematics Archives: http://archives.math.utk.edu/

Mathematics teaching material, software, publications and much more.

MathSearch: http://www.maths.usyd.edu.au:8000/MathSearch.html

A searchable index of 60,000 documents on mathematics and statistics.

This is Mega-Mathematics!: http://www.c3.lanl.gov/mega-math/

This is a friendly-looking site based at the Los Alamos national laboratory in the US. Mathematical games, knots and other features that make maths almost, but not quite, interesting.

Medicine

Doctors Guide to the Internet: http://www.pslgroup.com/docguide.htm

Not only a searchable index of Internet medical resources but also listings for new drugs, conferences and a bulletin board where doctors around the world can exchange messages. Has a regular update of new sites too. All in all this is just about the best site on the net for medical doctors. Let's hope it makes up for them just having a courtesy title eh?

Internet Resources for Pathology and Laboratory Medicine:
http://www.pds.med.umich.edu/users/AMP/Path_Resources.html

Internet resources for pathologists — everything from gynecologic pathology to veterinary pathology. Also has listings for jobs, conferences and much more. An excellent site.

MEDguide: http://www.medguide.net/

A searchable listing of medical Internet resources, plus a 'chat-room' for medics.

MedWeb: http://www.gen.emory.edu/MEDWEB/medweb.html

This is a searchable index of biomedical Internet resources, compiled by Emory University Health Sciences Center Library. Is is fairly comprehensive and has a sensible interface (i.e. not overloaded with graphics). In fact, it's so good that we can almost forgive it for succumbing to the fad of having a capital letter in the middle of its name. Almost.

Pedinfo: http://www.uab.edu/pedinfo/index.html

On-line information for paediatricians and others interested in child health. Organised into links for different medical specialities, institutions, software and many other useful resources.

Microbiology

Bugs in the News: http://falcon.cc.ukans.edu/~jbrown/bugs.html

This truly wonderful site is a collection of articles by Jack Brown of the University of Kansas. A range of microbiology related subjects are dealt with in layman's terms, with links for those interested in learning a bit more. Now why don't the media concentrate on telling the public about sites like this?

Bugs on the Web Index: http://bugs.uah.ualberta.ca/webbug/index.htm

This site has articles on various aspects of microbiology of interest to scientists in the field e.g. monitoring of water quality, enterovirus detection in cerebrospinal fluid by PCR. It also has links to other microbiology resources.

The Microbial Underground: http://www.gmw.ac.uk/~rhbm001/index.html

This is an excellent site which features an on-line course in medical microbiology, as well as links to medical, microbiological and molecular biology sites. A real labour of love which shows what one person (in this case Mark Pallen, a lecturer at Imperial College, London) can do. You will also find hypertext versions of four articles from the British Medical Journal introducing the Internet to medics. But why is it called The Microbial Underground? Visit the site and find out.

Molecular Biology

Cell and Molecular Biology On-line: http://www.taic.net/users/pmgannon/

A nicely arranged list of molecular biology databases, on-line tutorials, links to research groups, electronic publications and other useful things.

The Genome Database: http://gdbwww.gdb.org/

This site contains genomic mapping data from the various groups involved in the Human genome project. The information is made available to scientists everywhere via a search engine. The front-line of science available to you at your own desk.

Pedro's BioMolecular Research Tools:
http://www.public.iastate.edu/~pedro/research_tools.html

This is a real labour of love — a definitive list of molecular biology resources on the Internet, all annotated and rated. Here you'll find links to nucleic acid and protein sequence databases and their search engines, molecular biology guides and tutorials, on-line journals and much more.

If I wanted to convince a molecular biologist of the worth of the net, I would simply give them this URL and let them browse.

Molecular Biology Protocols: http://research.nwfsc.noaa.gov/protocols.html

An on-line set of molecular biology protocols brought to you by the Northwest Fisheries Science Center. You'll find detailed protocols for a range of molecular biology methods from DNA purification to sequencing.

PCR Jump Station: http://apollo.co.uk/a/pcr/

A compendium of polymerase chain reaction theory, protocols, software and much more. If you are a PCR user or are thinking of becoming one, add this site to your hotlist.

Neurosciences

Neurosciences on the Internet: http://www.lm.com/~nab/

This site aims to be a comprehensive set of links to neuroscience information on the Internet. Information on mailing lists, web sites, beginners guides and latest research are all here.

Ocean Sciences

USGS WR MCS Homepage: http://walrus.wr.usgs.gov/

The U.S. Geological Survey Western Region Marine and Coastal Surveys Home Page to give it its full title. Detailed information produced by the survey and links to other sites.

National Marine Fisheries Service: http://kingfish.ssp.nmfs.gov/

The NMFS administers the National Oceanic and Atmospheric Administration's programmes supporting the conservation and management of living marine resources. Find out all about the programmes here, including a copy of the annual Our Living Oceans Report.

Paleontology

The PaleoNet Pages: http://www.nhm.ac.uk/paleonet/

This is a linked system of mailing lists, WWW sites and other Internet resources designed to aid communication between professional palaeontologists. Palaeontology software, images discussions, job adverts and textual information. A definitive 'one-stop shop' for Palaeontologists.

Paleovision:
http://www.nhm.ac.uk/paleonet/PaleoVision/PaleoVision.html

Based at the Natural History Museum, this site is a project to make palaeontology data and collections available to scientists around the world. A good example of the latest technology being applied to the oldest science.

Physics

Astrophysics Data System: http://adsabs.harvard.edu/physics_service.html

A searchable database of 250,000 astronomical abstracts, 420,000 space instrumentation and engineering abstracts and 215,000 physics and geophysics abstracts.

Atomic Masses: http://isotopes.lbl.gov/isotopes/toimass.html

Experimental and theoretical atomic mass data.

Atomic Physics on the Internet: http://plasma-gate.weizmann.ac.il/API.html

A definitive collection of internet resources for scientists in the atomic physics field.

FD On-line: http://www.tfd.chalmers.se/CFD_Online/

A very good list of Internet resources for computational fluid dynamics.

Crystallography World Wide: http://www.unige.ch/crystal/w3vlc/crystal.index.html

A comprehensive collection of crystallography resources on the net; organisations, conferences, software, journals, databases and so on.

Fusion Power: http://www.fusion.org.uk/

An excellent site with an introduction to fusion power, more information, experimental data and links to other fusion resources. This site belongs to the United Kingdom Atomic Energy Authority — if only all public bodies had sites like this one.

General Relativity around the world:
http://jean-luc.ncsa.uiuc.edu/World/world.html

A collection of information, software and links concerned with Relativity. It's a great site but, oh dear, you just know who the server is named after don't you? At least it wasn't called nimoy.

High Energy Physics Information Center: http://www.hep.net/

Was about to host a video conferencing tutorial when we visited, which impressed us at least. High energy physics information, experiments and links.

High Energy Physics: http://www.cern.ch/Physics/HEP.html

A definitive set of links to High Energy Physics laboratories and resources around the world.

National Nuclear Data Centre: http://www.nndc.bnl.gov/

Provides information on neutron, charged particle, and photonuclear reactions, nuclear structure and decay data.

Nuclear Information WWW Server: http://nuke.handheld.com/

A comprehensive collection of links to Internet nuclear resources, power plants, research organisations, regulatory bodies and more.

OSA OpticsNet: http://www.osa.org/

This is the site of the Optical Society of America and has a comprehensive set of links to Internet resources relevant to optics and photonics research, applications and industry news.

Particle-Surface Resources on the Internet:
http://chaos.fullerton.edu/mhslinks.html

A definitive collection of Internet resources for scientists in the particle-surface field, maintained by Mark Shapiro at California State University, Fullerton. Has links to well over 100 databases, laboratories and other resources. Also has a jobs listing.

Physics news: http://www.het.brown.edu/news/index.html

News relevant to physicists. Culled from a number of sources.

Physics Resources: http://www-hep.phys.cmu.edu:8001/physics.html

A collection of physics resources on the net; conferences, jobs, journals, laboratories, software.

Plasma on the Internet: http://plasma-gate.weizmann.ac.il/PlasmaI.html

A definitive looking collection of links to plasma physics labs, databases, conferences, software and jobs.

SAM Project: http://aniara.gsfc.nasa.gov/sam/sam.html

The SAM (Systematic Accurate Multiconfiguration calculations) Project is a collaboration between atomic physics groups around the world to produce, collect and distribute accurate atomic data. The data and other information is available here.

The Physics of Superconductivity:
http://www.isisnet.com/MAX/science/physics/super.html

Superconductivity-related tutorials, images, applications, resources. Everything you might conceivably require on the subject. It also has a link to a chat room for research physicists called virtual Physics.

TIPTOP: http://www.tp.umu.se/TIPTOP/

The Internet Pilot TO Physics provides an excellent overview of physics resources around the world.

Waves: http://www.li.net/~stmarya/stm/links.htm

A collection of resources on waves: electromagnetic, sound, radio, infra-red, laser, UV, x-ray, gamma-ray.

X-ray WWW Server: http://xray.uu.se

From the University of Uppsala, Sweden comes this site full of X-ray information: the COREX database and bibliographies, Hencke X-ray scattering factors and much more data and resources.

Plant Science

Arabinet: http://weeds.mgh.harvard.edu/atlinks.html

A collection of all of the Arabadopsis information on the Internet. Everything from DNA sequences to where to get seeds.

The British Society for Plant Pathology Home Page: http://www.bspp.org.uk/

An information packed site about plant pathology: conferences, scholarships, publications, a pioneering on-line journal (Molecular Plant Pathology) and links to databases and other sites world-wide. Oh, and some information about the BSPP too. A wonderful, comprehensive source of information which should be on the hotlist of anyone interested in plant science.

Center for Aquatic Plants: http://aquatl.ifas.ufl.edu/

Photographs, line drawings, textual information and even video footage of all types of aquatic plants. Would you believe that there is also a link to the Prohibited Aquatic Plants List?

Internet Directory for Botany: http://herb.biol.uregina.ca/liu/bio/idb.shtml

A vast collection of plant science resources. The site comes in two flavours — organised alphabetically or organised into subject categories. Both are searchable. Links to information on everything from paleobotany to pollen.

A Survey of the Plant Kingdoms:
http://www.mancol.edu/science/biology/plants_new/intro/start.html

This beautiful, if graphics-heavy, site is a study of the diversity of the major plant groups. The information provided here is to student level but there are links to more advanced information elsewhere at each stage.

Psychology

Cognitive and Psychological Sciences on the Internet:
http://www-psych.stanford.edu/cogsci/

This is an index of research-oriented psychology and cognitive science links i.e. it does not cover practical aspects of mental health.

Internet Mental Health Resources:
http://www.med.nyu.edu/Psych/src.psych.html

A plain index of internet resources dealing with different types and aspects of mental illness.

PsychCrawler: http://www.psychcrawler.com/

This is a searchable database created by the American Psychological Association to provide access to high quality psychology Internet resources.

Psych Web: http://www.gasou.edu/psychweb/psychweb.htm

Psychology-related information for students and teachers plus links to resources elsewhere. The resources for students are particularly good (e.g. a ranking of US PhD programmes).

Sensation and Perception:
http://www.guam.net/home/bmarmie/sandp.html

This is a wonderful site containing links to tutorials on sensation and perception, collections of optical illusions and a collection of audio samples that demonstrate various aspects of auditory perception and its analysis. In short, it will appeal to both specialists and those with a general interest in the subject. Go on, visit it — you won't regret it.

Virology

All the Virology on the WWW:
http://www.tulane.edu/~dmsander/garryfavweb.html

This has been set up by the Garry Lab at Tulane University and, yes, it does live up to its name. There can't be many virology-related WWW sites that have been missed and they have been arranged in a sometimes inventive way. For example, links to images of viruses have been indexed as "The Big Picture Book of Viruses", a simple yet useful step that makes the most out of the WWW's ability to draw together disparate sources of information together. There are also links to on-line virology courses, techniques genome sequence data and much, much more.

WWW Server for Virology: http://www.bocklabs.wisc.edu/Welcome.html

Another comprehensive list of virology links, this time from the University of Wisconsin. Much the same resources as the Garry Lab site but arranged in a different way. Choose which one suits you best/is least busy.

Creating your own Web Site

Prof. Realitas **doesn't know it yet, but his department has been visited by a very keen secondary student, Lisa JiJeune. She surfed into the department's ad hoc home page this lunchtime, and looked at the sort of environmental research that's going on in the department. She was particularly pleased to see the statistical analysis of inhalers prescribed for asthma sufferers downwind of combi-fired power plants, work that had been posted by one of the department's enthusiastic lecturers. With all the energetic drive of youth, Lisa resolves to get the grades required by the department so that she can get her degree in environmental science. It would seem that the department has recruited another promising undergraduate to help fill its quota. If Professor Realitas had insisted on a www presence for each member of staff, and cogent and coherent expositions of their work, to say nothing of his own, what calibre of international students might the department be attracting?**

A surprising amount of information has been written about creating web sites, and much of it is useful. Not surprisingly, most of this information is available on the World Wide Web itself. Before you hie off in search of answers, though, (and we're going to suggest some useful places to start), do glance through this chapter to familiarise yourself with some of the basics of web site construction. We're seeking to cover the basic components of the Internet, after all, and as practicing scientists, with some personal experience in building useful sites, we may have something useful to say.

It may be worth reiterating what has come to seem the abiding principle of the Internet, in general. *Information wants to be free.* We would assume, if you wish to build a web site, that you expect to provide information to people who want to be informed. The particular audience you may seek to inform or educate, depends, of course, on the sort of information you wish to provide.

Perhaps the most important components of the successful web site, then, particularly for scientists, who naturally seek to communicate in the most precise manner possible, are: (i) simplicity, and (ii) straightforward ease of access to the information you wish either to provide, or possibly (iii) to retrieve. We're going to spend the remainder of this chapter reiterating these fundamental points, with models, pointers, and examples. We want, in the context of these points, to cover the following basic questions:

WHY BUILD A WEB SITE?

Because, in this geometrically expanding medium, lack of presence means that you, your project, your group, your department, your institution, company or school are not contemporary — worse, it will soon come to appear that you do not actually exist. We can't say it better than that, and so we won't say anymore; take it as given — you must have a WWW presence. This point will be so universally obvious by the publication date of this book that you will probably be wondering, while you read this, why we're even bothering to note it.

WHAT IS THE POINT OF YOUR OWN WEB SITE?

Do you want to provide information about: yourself, your project, your group? And how can the interested browser use this information? It is simply important to plan very carefully what the point of your communication, or information dissemination, is, and what the best way might be to present your information.

The best possible presentations are almost always the simplest ones. Probably the most important information you can provide is simply how an enquirer can contact you. The second point, perhaps, may be a simple description of what you actually do, or what your project is about. Now it's simple to create an abstract of your group's work, isn't it? Well, whether this abstract is presented on a sheet of paper, or in a prospectus or catalogue, or electronically, makes no difference, really, except that your colleagues, present and potential, will want to find you through the Internet, and they *can't* find (worse, won't bother even to look) that paper prospectus. They'll be searching for you, or someone with your sort of interests, on the WWW, and you *must* be there, or you will, in effect, not exist. But then we've said that already. And you agree, of course.

So let's start there: contact information, and a summary of what you do. Here's a simple, textual example of what we mean. We've already pretended we're Professor Realitas. Now let's pretend you're David George, nematologist extraordinaire. And why not? You painstakingly key in your details on your own friendly word-processor, and this is the text.

```
                              untitled

David W. George, PhD
Senior Lecturer
Department of Invertebrate Physiology
Red Brick University, Old Established City
Home County, United Kingdom, NE AR U
Phone: (0172) 12345
email: david.george@rbu.ac.uk


Research:

Our group is particularly interested in nematode physiology at
both individual and population level. That is, we are seeking to
determine how the nematode population as a whole reflects and
reacts to particular insults in the environment. Accordingly, in
our research trials at our field station in the Outer Hebrides our
measurements of soil chemistry and nematode populations, have
correlated with our measurement of individual nematode
reactions to particular toxins and pesticides.
```

So here's an example of a simple text describing a scientist, and, as it happens, *his* work. Too bad his work takes him to the Outer Hebrides, and not say, the Caribbean, but what the hay, maybe he just loves Celtic folk music!

Now, what our David wants to do is to put this text onto a WWW page, so that the information is available, to any browser contacting the server, like this:

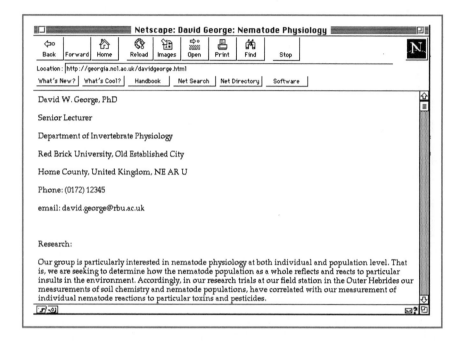

The translation process that our David will use, in order to get his text presented on a browser, is really very simple, but we can make it look quite complicated!

```
<HTML>
<HEAD>
    <TITLE>David George:   Nematode Physiology</TITLE>
    <X-SAS-WINDOW TOP=50 BOTTOM=451 LEFT=31 RIGHT=538>
</HEAD>
<BODY>
<P>David W. George, PhD</P>
<P>Senior Lecturer</P>
<P>Department of Invertebrate Physiology</P>
<P>Red Brick University, Old Established City</P>
<P>Home County, United Kingdom, NE AR U</P>
<P>Phone:  (0172) 12345</P>
<P>email:  david.george@rbu.ac.uk</P>
<P> </P>
<P>Research:</P>
<P>Our group is particularly interested in nematode physiology at
both individual and  population level.  That is, we are seeking to
determine how the nematode population as a whole reflects and reacts
to particular insults in the environment.   Accordingly, in our
research trials at our field station in the Outer Hebrides our
measurements of soil chemistry and nematode populations, have
correlated with our measurement of individual nematode reactions to
particular toxins and pesticides.  </P>

<P> </P>
</BODY>
</HTML>
```

This figure shows the basic 'html' code that David will need to use to display the text he wishes to present.

In the olden days, before html editors were very good, web authors used to do this simple coding by hand. And you can see that apart from the top and tail, it's just a matter of inserting <P> where you want a line break. If you use the above document as a template, it will work. But why do things the hard way?

Nowadays, html editors are really good, so it's a very simple matter of importing, copying and pasting, inserting (whatever particular method you or your editor likes) the text information into the html editor, and presto, all the codes are written for you. Usually html editors do like to receive ASCII text, but they'll probably be able to take the product of any word-processing package, within the nearest future. We'll tell you more about html editors — including where to get them — later in the chapter.

Now as a textual presence, you can't really fault our David. Maybe he could have used some boldness in some of the lines, on his name, and on the research title, for instance, but he's a modest sort of chap, and the information is there, and available to anyone who wants it. Note that David has used his professional, office or lab phone number. Why should anybody in the world need to get to your home number? To give your family hassle because you annoy nematodes? You don't need it, do you.

David might have included the simplest sort of direct contact capacity, otherwise known as the 'mailto' option, but he *has* included his email address, so that's fine, really. We'll discuss 'mailto' action below, when we consider retrieval of information from the browser.

It could be that having accomplished this simple task of preparing a text, translating into html code, and mounting the file on a web server (about which we have something more to say after we discuss some fancier variations of building a web site), our David is perfectly satisfied. And we say, fine. Good for you, David, you're here, and that's what counts. End of story.

It may be, however, that David's group has just generated some great new data, and he really would like it to be available for foreign browsers. Now what should he do? Why follow his nose, of course, and continue to build a lovely web site upon these sturdy foundations.

WHAT IS THE BEST WAY OF BUILDING YOUR SITE UPON THESE FOUNDATIONS?

Well, what are the other possible components of a web site? We've mentioned text, and to be honest, except for diagrams and graphs of data, you can't get much better dissemination of information than through the printed word. Why do we publish that way, if this is not true? So, as we've gone to some lengths above to illustrate, you'll need to get the text right, as your silent representative or constant presence on the net. But of course publications include images which help to demonstrate the point of the text, or to supplement the message. Typically, these images can be either graphs, photographs, diagrams or sketches.

Images:

The World Wide Web, of course, unlike its 'gopher' antecedent, handles graphic images with ease, though these images must be in an appropriate format for the electronic browsers 'out there' to read, interpret, and then display on their human's monitor.

Images received by browsers on the WWW come in two kinds: gif or jpg (jpeg). These subscripts identify processed files that typically are dramatically compressed compared with bitmaps (filename.bmp, as rendered for example by the Windows Paintbox) or PICTs (from macs), being up to 90% smaller than the original. Now it's worth remembering (and we won't forget to remind you again along the way!) that in planning a typical page or single image the web author should aim for a maximum of 50 or 60 kb.

But a single useful size photo with reasonable resolution will easily take up that sort of requirement with no problem.

So when you add pictures, or images, to your page, you immediately set up a conundrum for the person reading it. Is the picture really worth a thousand words to them? A thousand words will come screaming down the wires (figuring 10 bytes/word so about 10kb per 1000 word document) a whole lot faster (well, about 5 to 10 times faster, that is, you get the whole file in one fifth or one tenth the time — of course, transmission speeds are the same) than a 50–100kb image. That won't matter much if the browser perusing your presentation is on the same continent as your server (unless they have a slow modem feed, when it will matter a great deal), but trans-Atlantic or trans-Pacific links can get pretty congested, and browers will quickly lose patience if your picture is only really worth a hundred words, say.

So it turns out that good old text is pretty damn good information, after all. Still going strong, all these years after Gutenberg, or should we go back to the Rosetta stone for early textual citations? Anyway. We don't mean to debunk images on the WWW, far from it. We just want you to be sure that your pictures do the job they're supposed to do. And we'll talk about thumbnails as well, which can provide useful information that can be enhanced at a click on the foreign browser's whim.

We're quite sure that the scientists we're writing to will have seen movies, cartoons, and even three-dimensional images at some point in their lives. Is the WWW the place for scientists to worry about presenting or developing these kinds of images? Well that depends a little on your science, doesn't it. If you makes your living building 3 dimensional images of nematodes, then you'll probably want to present those images. Or if your work involves writhing, 3-D nematodes doing what nematodes do with each other, then you're also likely to want to show them at it. We think, however, that if your work is involved in these areas, you're the sort of person who already knows how to get the video images into an appropriate 3-D format, so that 3-D viewer plug-ins in foreign browsers can translate and exhibit the virtual nematode. That's just fine.

But most of us are basically concerned with text and two dimensional images. That's the sort of presentational world we live in. So let's do a brief run-down, just as we did above with text, on how actually you can get your images into the appropriate gif or jpg format for WWW presentation.

There are really three ways to get the image you want onto the screen on a WWW browser: (1) Software produced images; (2) scanned images; (3) screen-capture images. Okay; (4) digitally produced photographs.

(1) The software you might use to produce your graphic presentations (like Corel Draw, or Claris Impact), or to draw your data graphs (like Cricket Graph, FigP, Excel, or DeltaGraph) can be directed to save your image in an image file (like .bmp, or .pict). Here's an example of our David's dramatic new data indicating that, within a very narrow dose range, PCBs actually enhance the size of individual nematodes. David has put his data into his data graphing software.

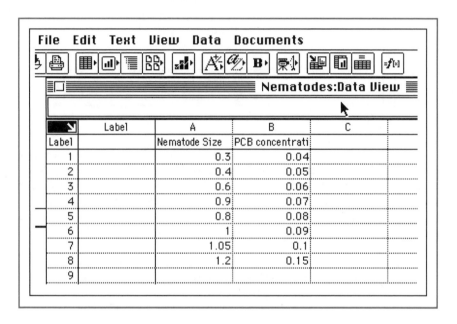

Now, he asks his graphics software to plot the data.

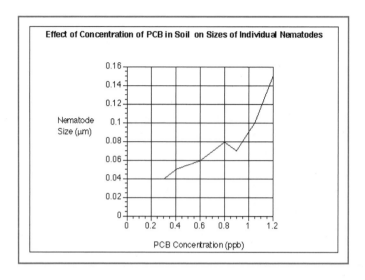

Very nice David, though I'm not sure I do believe it. Have you got osmolarity comparisons of your different soil samples? But anyway. Now that the thing is plotted nicely, David seeks to export the plot into a useful file which he can manipulate for presentation on the WWW. After 'selecting all' the components of the image he wants to export, he exports the information into a PICT file, like this:

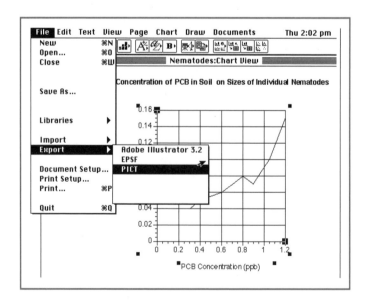

Newer versions of these graphics applications might even include the capacity for directly saving as a .gif or .jpg file. If not, have no fear, however, for there are graphic conversion applications (Graphic Converter for Macintosh, and Paint Shop Pro for PC-users) which will open any image file, and convert it into a WWW-friendly format. We've included a friendly little table, nearby, with FTP sites for acquiring this extremely useful shareware.

TABLE FOR FTP SITES OF GRAPHIC CONVERTER FOR MACS AND EQUIVALENT FOR PCS

Graphic Converters for Macintosh Computers

mic2-atm.lancs.ac.uk
  /mirrors/info-mac/_Graphic_&_Sound_Tool/_Graphic/graphic-converter-242-fr.hqx
ftp.doc.ic.ac.uk
  /Mirrors/sumex-aim.stanford.edu/info-mac/gst/grf/graphic-converter-242-fr.hqx.gz

  ftp://ftp.info.au /micros/mac/info-mac/_Graphic_and_Sound_Tool/_Graphic/

ftp.uoknor.edu /mirrors/info-mac/gst/grf/graphic-converter-242-de.hqx
sumex-aim.stanford.edu /info-mac/gst/grf/graphic-converter-242-fr.hqx
ftp.hawaii.edu /mirrors/info-mac/gst/grf/graphic-converter-242-fr.hqx
ftp.uoknor.edu /mirrors/info-mac/gst/grf/graphic-converter-242-fr.hqx
ftp.maricopa.edu /pub/mac/Graphic_and_Sound_Tool/graphic/graphic-converter-242-fr.hqx
ftp.cso.uiuc.edu /pub/systems/mac/info-mac/gst/grf/graphic-converter-242-fr.hqx
ftp.wustl.edu /systems/mac/info-mac/gst/grf/graphic-converter-242-fr.hqx

PC: **Paint Shop Pro**

How to get Paint Shop Pro:

Both psp311.zip and psp32.zip can be downloaded from JASC's home page

http://www.jasc.com/pspdl.html

or from mirror sites listed there. In the U.S.A., the compressed file can also be downloaded from http://www.winsite.com/info/pc/win95/desktop/psp311.zip.

Mirror sites around the world can be found through the Oakland Software Respository's Virtual SoftwareLibrary:
http://castor.acs.oakland.edu/cgi-bin/vsl-front/QuickForm
  by searching for keywords "Paint" and "Shop"

And finally, here's what our David's graph would look like, live on the WWW, now that he's converted it to a .gif file, and has loaded it onto his server, and linked it as described below, so that a foreign browser can come in and find it, as I have just done:

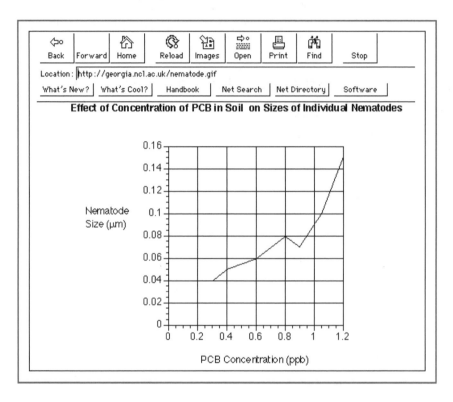

It might be worthwhile, here, just to note that image presentation on the WWW is one remarkable way in which that medium is far ahead of hard copy publishing, since colour presentation is not a problem at all. We're assuming everyone is using colour monitors, perhaps a rash assumption, but surely correct for at least 99% of all users. So these images, of course, on the WWW are all in naturally glowing, pristine colour.

(2)   If you're scanning images (photographs, usually) on a flat-bed scanner, you'll probably be using something like a full PhotoShop 3.0 to acquire the image, which does include the capacity to save as a .gif or .jpg file. You have to work with the software and the image yourself, of course, until you get a useful size (in kb, we mean), of good enough resolution

(100-150 dpi is perfectly adequate, we think, for most monitors, and so is 256 colours) and brightness (for some reason, we've found that it's useful to get about a 50% enhancement in brightness on scanned photographs, with additional contrast to taste). If you save the image in the WWW-friendly format, then it's ready as-is to put onto your WWW server. Otherwise, use your friendly graphic converter software to create the right file format. And that's it, as far as scanned images goes.

Here's a scanned image of my favourite pet, just for show:

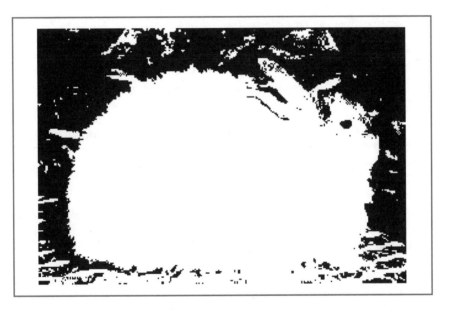

(3)   Screen capture software is a really handy sort of tool, and we're familiar particularly with "Flash-It" for Macintosh machines. With this extension installed, you merely position the cursor out of the way, hold down the shift, the command=apple key, and the '3' key simultaneously, and, although nothing happens immediately, you then get a movable box linked to the depressed mouse button which with a little practice you can position around just that area of the screen that you want to 'photograph'. Release of the mouse button elicits a satisfying camera shutter sound, and you're provided with the option of saving the PICT file you've just made in an appropriate folder. All ready for translation into a .jpg file with your handy graphic converter, right! The screen shots for this book were all produced in this way.

TABLE OF SITES TO FIND SCREEN CAPTURE SHAREWARE

WinCopy for PC-users

http://users.aol.com/informatik/wincpy.zip

or, if this site is busy, or access is limited, try
http://www.shareware.com
(search for the keyword wincpy in the MS-windows (all) category)

Flash-It for Macs

mic2-atm.lancs.ac.uk /mirrors/info-mac/gst/grf/flash-it-302-de.hqx

ftp.doc.ic.ac.uk
/Mirrors/sumex-aim.stanford.edu/info-mac/gst/grf/flash-it-302.hqx.gz

ftp.doc.ic.ac.uk /Mirrors/ftp.cnidr.org/pub/NIDR.tools/Mac/Dir-Graphics/Flash-It_3.0.2.bin.gz

(4)   Digitally produced photographs. Okay, fancy-pants, you've got yourself a digital camera. So you can by-pass the scanning step, and download the PICT or .bmp image directly into your computer. Give it a quick conversion to .gif or .jpg, and you're away.

We said that we'd mention 'thumbnails' again, and this might be the place to do it, now that we've taken care of all the ways to process images. Using the graphic converter application, you can size the image to whatever size you want, on the screen. My (LW) experience, however, though I'm probably using the software wrongly, is that the image stays the same size, for WWW purposes, as the original. (That's pretty damn frustrating) So I find it extremely handy to size as I want on the monitor screen, and then Flash-It, for saving in an appropriate size, whether thumbnail as below, or full view, as above.

Here's that favourite pet again, in thumbnail version. Pretty simple, eh? For my next trick . . .

Do you Need Sound on your WWW page?

We doubt it very much, unless sound is your science. The particular hiss of a Galapagos tortoise, or the whine of a rare cicada may be what you want to capture and present. Well good, and the presentation is simple too. Digital sound files are usually called .au files, and if you can get them recorded, and presented in this format, then somebody's browser will be able to download them, and activate the appropriate application to play the files. On the Macintosh, I (LW) can record directly into .au files, so it's no problem for me. On the other hand, I only present sound files of the folk music (Northumbrian Ceilidh Band) I have on my own server, but there's just no sound in the science I serve. Sound files, to get any useful snippet of time, tend to be pretty big and unwieldy, and really carry so little information, for non-specialist purposes, that there just isn't much point.

Presenting Tables on the WWW:

Until very recently, tables were an html nightmare. And that was a blow; lots of data, after all, look best in tables. But now they're really quite simple to prepare, because the html editors have figured out how to do it. At least the Macintosh ones, like Claris Home Page, understand how.

David has opened the html editor application, and he is trying to figure out what the basic layout of his table should be. Well why not present the data as they appear in the spread-sheet, then? Okay, so he'll need two columns:

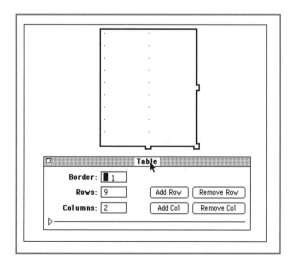

Now carefully, David keys in the data into the individual slots. Laborious this, but not as bad as sizing those wee worms, now is it?

Nematode Size	PCB Concentration
0.3	0.04
0.4	0.05
0.6	0.06
0.9	0.07
0.8	0.08
1.0	0.09
1.05	0.10
1.2	0.15

There, it couldn't be easier. If he likes, David can centre the data in the individual cells. But maybe he prefers the standard left justification, and why not! We'd hate to scare you by showing you the html code for this wee table, and so we won't. Remembering our car analogy, we're not going to discuss the mechanics of piston thrust either, okay?

Wrap-Up Section on Web Page Style and Presentation:

We've indicated that we think that for any scientific web page, simplicity is the best policy. And we think we've made our preference for text pretty clear, though we do recognise the usefulness of pictures in one's presentation. Where would we scientists be without graphs and tables, anyway?

But there are useful ways to expedite the browser's direct information acquisition that should be mentioned, somewhere, and so in this wrap-up we'll seek to do just that. Directories: some people hate 'em, some love them. But you don't want to lay everything on the browser at the front door, really, so it can be useful to provide at least an initial entry point to your information resource pages, through which the browser can make their own select choice of subject matter. This sort of entry is also very handy when you begin to accumulate bits of information that might not all fit together, necessarily. Our David might suddenly develop an interest in monoclonal antibodies directed against some component of nematodes, which doesn't quite fit in with his main interest in how they grow. Or maybe it does, but our point is that if he includes a hypertext-linked directory on his front page, browsers interested in the one topic can go there, while others can go to the growth topic. It can be incredibly handy for the browser if you can include all the basic directory routing information in an initial screen shot, that they download on first entering your site. You can combine the initial directory within a table format, which makes simple comprehension easier, and removes the need for the browser to 'arrow' down through countless lines of text, to get to the hypertext link they want.

So a browser entering my home page can move directly into whichever area of my presentation they wish; the "Professional Activities" link takes them directly to our lovely immunodiagnostic data. Of couse, they can go directly to that area as well, bypassing the home page entry, and many do, merely by specifying the appropriate file at the end of the URL. But the initial directory access information arrives quickly on my 'default' home page, before even the 'hat' images, and particularly before the background 75k image, which arrives only after the browser has been studying the 'lay of the land' for some time. So it's a simple matter for the browser to move along, once they see how the whole set of pages are laid out.

Okay, then, directories are fine; people do hate 'em when they have to click or page through several sets of directories to get to the real 'content' of the site. If you've been surfing around the net, you'll know, by now, what we mean when we say that this can be a real hassle, and can contribute to extreme frustration, when you're just trying to get the damn info! So it's always nice to provide your hypertext links in a manner that conveys information directly, and also provides opportunities for enhanced learning too.

Another important point, which keeps popping up in reviews of web sites, is still the vexed issue of large in-line images. Image-maps, which we haven't dealt with at all, in this chapter, particularly because they can be unwieldy, and must be entirely downloaded before you can click on a hotspot, can be a case in point. Who needs them? Small image maps, we agree, have a utility, but we don't want to be waiting on a trans-Atlantic access, for long minutes and minutes waiting for a big image to get here in its dribs and drabs. Keep your information punchy, direct, and to the point, and you can't go far wrong.

Since we're expecting a reasonable European readership, we must note the wonderful service provided for free by the good folks at: http://www.systransoft.com/translate.html who will translate a WWW document (you specify the URL at which the original sits, and somehow they provide you a European language translation in a couple of minutes) into French, German, Italian, Spanish, or Portuguese, or back again.

Well, that's a wee primer on style, and presentation. Doubtless there's a lot more tips and useful information, but that's partly for you to go out and understand for yourself, using our handy table, at the end of this chapter, perhaps, of web style pointers.

HTML EDITORS AND SERVERS:

More notes on composition editors, confirming your presentation, and putting your html files onto the server for 'serving' to incoming browsers.

Contemporary html Editors:

Have we mentioned that 'html' stands for 'Hyper-Text Markup Language' yet? If not, sorry. Now there are excellent html editors for either PC or Mac platforms, and as we've mentioned above, they really do take the headache, such as it was, out of text-based html code. You have to choose the one that's best for your own machine, or your own tastes and budget, out of the list presented in the table below. We ourselves (well, LW, who's writing this section) are particularly partial to Claris Home Page on the Macintosh platform. Claris Home Page has been available, during its beta testing period, as freeware, but was launched as a commercial package during the autumn of 1996. The following list of editors can usually be found for any or either (Mac or PC) platform.

TABLE OF HTML EDITORS FOR MACS AND PCS

Claris Home Page 1.0 is currently available through authorized resellers at an estimated retail price of $99 and estimated education retail price of $79. Claris Home Page 2.0 has just been released.
Evaluation versions (1 month expiry freeware) are available from the Claris site: http://www.claris.com

Adobe PageMill (1.0.2 and 2.0)
http://www.telteksys.com/pagemill2.0.html

Here, why don't you (yes, that's you, the one doing all this easy-peasy reading!) do some search work for a change! To find the editors listed below, on your friendly WWW search engine, type in their names, using quotation marks:

HoTMetaL Pro
Tapestry
GNNPress
Microsoft FrontPage
Netscape Navigator Gold
Corel Web.Designer 2.0
Quarterdeck WebAuthor
DeltaPoint QuickSite
InContext Spider

Alternatively, a comprehensive Web Design resource service listing many editors for all platforms, is available at: http://web.canlink.com/webdesign/nl.htm

When composing your html files, it can be very handy to confirm, offline, just what they'll eventually look like to a foreign browser, when they eventually get loaded up on a server. Fortunately, www browsers are easily able to handle this challenge — you merely have to choose the file you're wanting to view, as opposed to keying in a URL, or hitting a hot-linked anchor on somebody's www page. We've included a wee snapshot of someone doing just that, below:

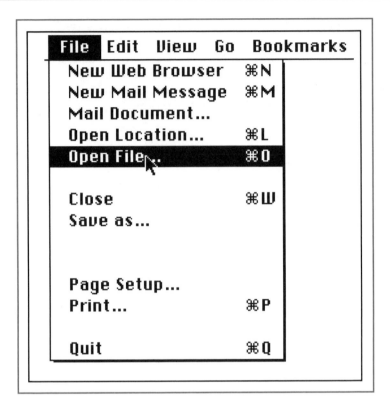

HTTP SERVERS:

These handy little application programmes are the workhorses of your personal serving system. Of course, if you're uploading your material (html files and images, for example) onto a server in another place (eg, your institution, or organisation's computer, or a commercial web-space provider), you won't have to worry at all about these programmes. On the other hand, if you want to serve browsers from your own ethernetted personal computer sitting just there on your desk, or in your physical department, then you'll need one of these guys.

In fact, there's nothing magical about these applications; once initiated, they accept incoming queries, and send back the files requested, if they're available. When you download one of these freeware programmes, you'll read the accompanying instructions, and typically will have to make one or two minor modifications to fit your file names. For example, serving from MacHTTP2.2, I changed the name of the default file, so that when my home page at http://georgia.ncl.ac.uk is accessed, the browser is directed to the 'home.html' file. That was about it, really, except to ensure that the application programme sits in the same folder or directory as your html files or images that it will be serving.

TABLE OF HTTP SERVERS

A list of useful shareware servers can be found at:

http://www.schnoggo.com/guideServers.html

which includes, in addition to commercially available servers:

Apache HTTP Server Project. Enhanced version of NCSA HTTPd. (UNIX)
http://www.apache.org/

MacHTTP | WebSTAR Home Page Full featured web server for Macintosh.
http://www.biap.com/

NCSA HTTPd UNIX-based Web server.
http://hoohoo.ncsa.uiuc.edu/docs/Overview.html

CERN httpd UNIX-based Web server with authentication and encryption.
http://www.w3.org/hypertext/WWW/Daemon/Status.html

EMWACS HTTP Freeware web server for WindowsNT.
http://emwac.ed.ac.uk/html/internet_toolchest/https/contents.htm

Jigsaw A new server from the W3C, written entirely in Java.
http://www.w3.org/pub/WWW/Jigsaw/

HOW CAN YOU ENSURE THAT 'THE WORLD' KNOWS ABOUT YOUR WEB SITE?

You'll have been defining your expected audience as you composed your own presentation. Since our audience in this book, we expect, is made up of other scientists who may have felt somewhat left behind by the Internet revolution, we'd bet that our readers will want, primarily, to be accessible to other scientists, and lay-people who are particularly interested in their own field. So how do you let these members of your community know (a) that you are there, with a www presence; and (b) where *there* is?

The Internet Search Engines — Adding Your URL

Well, first of all, you'll want to confirm that you yourself can find your own presentation. You want to know that the URL you'll be using, and providing as an access address, actually works. So check it out with your browser, but just to be sure, check it out with somebody else's browser. It can be very useful to have a friend somewhere else in the world, who you trust to give you an honest opinion of your site, and presentation. This friend can tell you whether the site is actually functioning, and be a first line of confirmation that it is presentable!

Having confirmed that your site is really live, and accessible on the WWW, you're now ready to be, as it were, 'hit' on by foreign browsers. And how will they learn of your site? Well, in a general sense, you might assume that international, or world-wide browsers, will use the same internet searching tools that you'd use. You'd go directly to your favourite search engine, wouldn't you, if for example you wanted to find out about nematode growth. If *your* research group is working on nematodes, then, you'll want to ensure that the search engines know *your* own URL, and useful key words or synopses of your project(s), just like they could identify for you the work of other colleagues around the world.

We've listed the biggest and best www search engines in the first part of this chapter. But to make things easier for you, there are what you might think of as composite sites which you can use to post details of your own presence (your own URL), such that you don't have to actually contact each of the search engines in turn. We've included both composite sites, and individual sites, in the table below. Depending on the backlog of URLs waiting to be catalogued, you might have to wait several weeks before you will be able to confirm, with a search of your own, that your site is actually listed on the search engine's books. And if you can find yourself, then it's a fair bet that other like-minded people, keying in the appropriate search terms, will also be able to find your site.

TABLE OF PROMOTIONAL SITES FOR PUBLICISING YOUR WEB SITE

as described by Lon Koenig at http://www.schnoggo.com/guidePromotion.html

Submit it! Easily get in all the major search engines
http://www.submit-it.com/

Entity Global Site Submission announces your presence to over 100 sites.
http://www.i-studio.com/entity

Pointers-to-Pointers Allows posting to special-interest sites and newsgroups.
http://www.homecom/global/pointers.html

Promote It! Lists good places to tell about your new site.
http://www.cam.org/~psarena/promote-it.html

But you might want to take it a bit slowly, and just key in your URL to your favourite search engine, and leave it at that.

Here's an example of the sort of posting you might make, to a search engine like Lycos, of let's say our David's minimum presence URL:

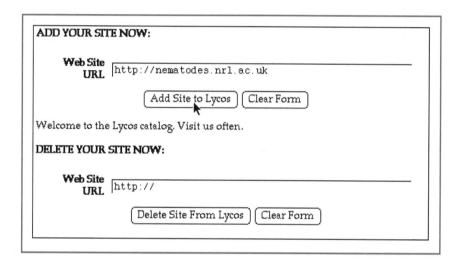

Once you're registered on the search engines, however, you've got to be prepared to be probed by robots, who regularly check your 'live' presence and constantly update their own database of active sites. That's in their best interest, of course — they won't want to keep a record of defunct sites, to send interested searchers haring off into the void, or unidentified server territory. Robot probing is usually very gentle; once in a while you might get 'web-whacked', but that's another story perhaps best left to develop into a full-blown urban myth.

You can pay, in these enlightened times(?), to have your home page information submitted and constantly updated, to dozens and dozens of search engines:

●CHOOSE 50 or 100 DIRECTORIES AND SEARCH ENGINES

There are over a thousand directories on the Internet *today* Our Internet Specialists weed through these directories on a daily basis to insure that you are getting the best quality possible. On a normal week we change our submission list by approximately 5%. We do this to ensure that our clients are getting listed on the directories and search engines that have the most traffic and widest appeal.

50 Directories and Search Engines Just $110.00

100 Directories and Search Engines Just $195.00 (SAVE $25)

(10% additional discount available in exchange for a link to our page.)

Once you are on these directories, you'll see an increase in traffic that you probably never even dared to hope for.

Yes, yes, and we've also heard certain scientists we do not deign to mention here referred to as snake-oil salesmen. We'd not have thought you would want to pay for publicity, even if your publications have to be labelled 'advertisement' to comply with U.S. federal law!

Or you can submit your URL to several search engines, for free, using one handy form, at the Submit It! site:

The Submit It! Submission Form

After you submit the completed form below, Submit It! will return a page with submission buttons for all the sites you selected. All the information you enter below will be encoded in that page so you won't have to retype it. Use that page to register your site with the catalogs and choose a category or categories for your site when necessary.

☒ Yahoo ☒ Whatsnew on the Internet ☒ Infoseek ☒ WebCrawler
☒ Apollo ☒ Starting Point ☒ ComFind ☒ InfoSpace ☒ Yellow Pages Online
☒ What's New Too! ☒ METROSCOPE ☒ LinkStar ☒ BizWiz ☒ WebDirect!
☒ New Rider's WWW YP ☒ Nerd World Media ☒ Alta Vista
☒ Mallpark

Title of your site:

Nematode Growth and Development

URL:

http://nematodes.nrl.ac.uk

Announcing your WWW site in the Usenet Newsgroups:

You've been spending a wee bit of time in a scientifically relevant newsgroup, perhaps, and every once in a while you see a posting from somebody to the effect that a new www site is up. Just because you've created a web site, however, doesn't necessarily mean that people will, or even should, want to visit it. Perhaps like our David above, you've created your site only as a 'calling card' presence, or an abstract of your group's work. Let's be honest, you don't have to alert the whole newsgroup to a mere calling card presence. Indeed, you can do this sort of alert, much more effectively, by including the address of the site in your signature file, which you can use when posting messages of your own to the newsgroup, or in individually addressed email messages.

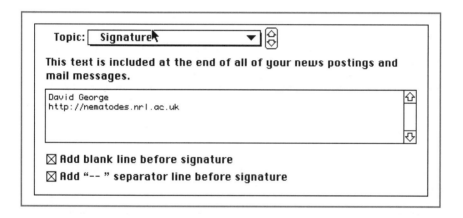

On the other hand, suppose you've created a web site with rather more than just a minimal presence. Suppose you're presenting a lot of data on how nematodes grow, on the effect of various toxins, etc., and you've got quite a story on your own project. This sort of expanded presentation could be useful to know about, as a resource of information (!) so you might want to make a wee public notice to that effect. The problem that you'll be faced with, once you've informed the interested world that your resource exists, is whether you can cope with the possible onslaught of browsers 'hitting' on your poor server. Depending on the subject matter, and timeliness of the resource, and the relative size of the interested audience, you could easily draw significant, 4 digit numbers of browsers to your site. Can your server handle the attention? The big institutional ones, or the commercial web space providers are equipped, usually, to cope with demand, but your little desktop server might become over-whelmed. So the question of browser demand is an important one, when you're deciding whether to serve from your desk, or from your organisation's bigger server.

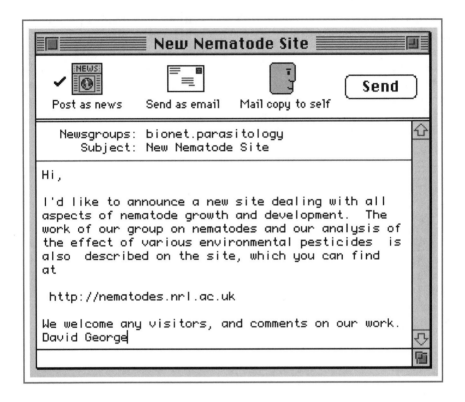

Announcing your WWW Presence in Dedicated Mailing Lists

A similar set of rules, or code of conduct, applies in mailing lists as applies in Newsgroups, as we've discussed in the appropriate chapter above. The difference is that mailing lists are rather more likely to be read by rather more directly interested people. So when it comes to announcing the existence of your www site, you can apply the same sort of logic we've used above for Newsgroup posting, viz. signature file for a calling card presence, or even as a regular reminder of the resources at your active www site; and/or a wee announcement of the really useful resources available at your new site.

Just an announcement, however, doesn't necessarily guarantee much of an audience. The audience, much like you yourself, actually, are particularly drawn to sites by testimonials, or word of mouth (email). Here's an example of what we mean. Some two years ago, I (LW) posted news of my developing web site, in which I present some of our immunodiagnostic assay data, in the MEDLAB-L subscription list. A few interested parties, as I can tell from the log my HTTP server keeps, visited the site, and looked at a few files. But the real fun began when the owner of the list visited my site, and then posted a testimonial about it.

The figure below shows the number of accesses to the URL http:// georgia.ncl.ac.uk which occurred after various announcements of the type we're discussing in this chapter. An announcement, or testimonial, typically elicits a large number of accesses, which die away after the first flush of enthusiasm. Thereafter, unless the site has very useful resources to offer, (and with the advent of the incredibly powerful search engines, lists of useful links are less and less important) accesses to the site return to a lower level.

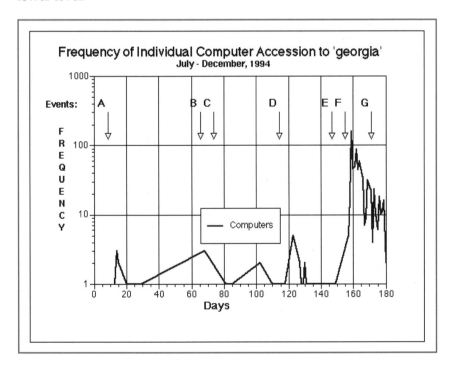

The accompanying tables provide an anecdotal history of events which I've compared with frequency of foreign computer (browser) access to the 'georgia' home pages, in a study of what brings people to one's home pages. I analysed accesses to 'georgia' in this study I conducted over 2 years ago. At the time, the idea of useful links to various other interesting sites was of pretty fundamental importance, since incoming new users didn't have a clue where to look for useful resources. This use of home pages is decreasing, while the importance of specific 'content' is increasing. Both components of one's home page, however, are still important, it's just that as we all get more sophisticated, we know more about *where* to look, and we really want to just, as it were, *find*.

A Brief History of the 'georgia' Home Page

Purchase PowerMac/PC March, 1994

Ethernet connection a couple weeks later

Quick access to 'receiving' net tools — fetch, telnet, Mosaic, newswatcher through University computing service.

April, May, June spent browsing, dealing with persisent memory problem (browser crash after only a few inline images). But newer versions of MacWeb are pretty fast!

Subscriptions to various listservers (MEDLAB-L, diagnost@net.bio.net)

Realising other institutions are rushing to set up their home pages, query University computing service about setting one up for the department.

Access net broadcaster/server software from University computing service.

A. Rudimentary home page set up on Georgia by 13th July. Placed a few images, and set up a few anchors to places of interest. Checked out by University computing service, who recommended minor modifications. Sit back and wait for the connection log to run up. Surprise, surprise: nobody's interested! But nobody knows I'm here!
URL (Uniform Resource Locator) goes on email signature — still very little response!

B. Include URL (iD signature) in note to MEDLAB subscription list — a small flurry of interest.

The Guardian (UK national newspaper) piece in OnLine appears (October 20, 1994) on personal home pages. (Thinks: ah ha, I'm not the only one into this!) Email The Guardian (online@guardian.co.uk) with comments.

C. Join up with Enrique Canessa's Who's OnLine — I've got to get on some list or other!

D. My letter on MacWeb sourcing of other people's home pages (it helps to see how somebody else has written up their html document, and some browsers will return the 'source' that the server is using) appears next week (October 27, 1994), with my URL included on my email address signature, in The Guardian OnLine, and elicits some browsing interest (minor).

Construct an Online seminar on the Internet for my department (Response: But what can the Internet do for us? — Thinks: But what can we do for the Internet?)

E. Indicate availability of Internet access to interesting biomedical sites, through my home page to MEDLAB subscription list (of which, by then, some 300 members); a few responses.

F. Rave review from Pat Letendre, who owns the MEDLAB-L list. The deluge!!!!

G. Note to Diagnost@net.bio.net, re home pages as promotional tool. Note elicits another respectable wave of interest.

Well, well, it's hard to believe that old 'georgia' has been going for over 2 years now! Anyway, the point of this section is that the sort of *status quo* of accesses which replaces the floods of enthusiastic browsers after any particular announcement, then, could be a result of the regular reminder of your site that you've included in your signature, as you post messages around the world, or it could be from the search engine listings, but it's just as likely that you'll be getting most of your regular traffic as a result of the interdependent network of mutual links around the world. Net surfing has not yet gone out of fashion, we guess, though we do think that the enthusiasm for spending hours at your computer flitting hither and thither, will wane. We think that people will be visiting web sites for a real purpose, and they'll be pretty offended if what they find is not content, but only links to somebody else's content.

Mutual Links with Other like-minded Sites

It still makes sense, however, not only to external browsers, but also for your own convenience, to have directly relevant sites linked through your own site. You could link these useful sites as a 'hot list' arranged in your own sensible prose, perhaps, or even in a directory-type format, if that suits. Then when you want to use a particular resource, you'll always know where you can find it quickly, through your own collection of links. We ourselves, for example, have composed web pages describing a couple of the so-called 'orphan diseases', in which the rare diseases we're particularly interested in are linked to various 'compendium' sites. Consider KO'D's site (http://www.compulink.co.uk/~members/), set up on a commercial server, which, in addition to some very useful content on Pompe's Disease, also includes very useful links which can help the browser to a better understanding of what is also called Glycogen Storage Disease.

Other GSD-related Resources on the Internet

● OMIM On-line Mendelian Inheritance in Man is a searchable database on all of t diseases. A lot of information but very technical.

● There is a database of GSDII mutations at the GSD II page of Dr Arnold Reuser's University. NEW

● More Pompe's information at the Glycogen Storage Disease Type II (Infantile) site metabolic pathway. NEW

● GSD I information from the American Liver Foundation

● The Muscular Dystrophy Association in Australia has information about severa

● Diseases of the Liver has information about the liver-related forms of GSD.

● This example of on-line diagnosis by a medical student is a good illustration of t face in diagnosing rare diseases like the GSDs.

● Rare Genetic Diseases In Children: An Internet Resource Gateway is an excellent metabolic diseases - the comprehensive resource section alone is well worth a vis one of its features. NEW

They're the ones that appear as underlined text.

The Glycogen Storage Disease site, as well as say, Familial Hypophosphataemia, in which LW is particularly interested, has mutual links back and forth to the Rare Genetic Diseases web site.

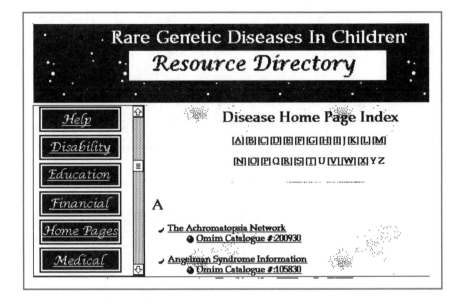

Interested browsers, on finding the Rare Genetic Diseases site, can naturally follow their links to the disease they're particularly interested in. That makes sense. Similar links for our David's nematodes, for example, should be pretty obvious to nematologists who have spent some time browsing in their own discipline. We'll make more of a meal of these mutual links in our chapter on the Tea Break, coming soon to a common room near you.

Now in this chapter on website composition, it would be pretty remiss of us not to mention how to place URL tags. Using the old, textbased html codes, it was very easy — you just put in the correct punctuation. Let's say you want to create a link to the American Type Culture Collection, so that browsers can click their way over to the ATCC.

Your prose is moving along nicely . . . 'another very useful place to check out the possibility of available hybridomas is the <a href="gopher:// gopher.atcc.org>American Type Culture Collection</a>, which can supply frozen cells to virtually anywhere in the world.'

This paragraph would appear to the browser merely as: 'another very useful place to check out the possibility of available hybridomas is the American Type Culture Collection, which can supply frozen cells to virtually anywhere in the world.' In which the underlined words are hotlinked hypertext, as you've specified with your html punctuation. You've just set your first URL. If you happen to be using an html text editor, there will be simple icons, or menu driven settings which you can select, having highlighted the text or image that you want to link to a URL. Here's a wee example of the same thing, as developed on the Claris Home Page html link editor:

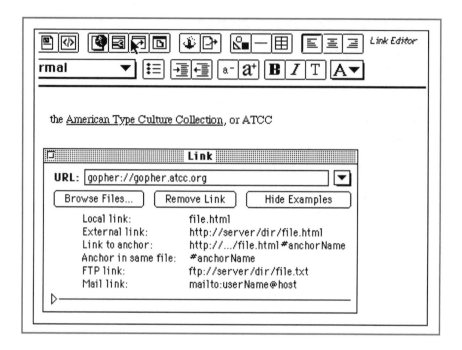

Internet Electronic Magazines and Hot (or is that Cool?) Site Lists:

Let's suppose your site is really incredibly fabulous — maybe it has resources (the fabled 'content' that is constantly discussed in newsgroups about the www) on say, dolphins, that are just so useful, and enlightening, and exciting, that an electronic digest like Internet Digest checks it out, and then distributes information about how wonderful your site is to its vast mailing list. With an announcement like that, you can expect a serious deluge of interested browsers, anxious to see what you've got to say. You could be featured in the 'Top 5%', which earns your site the right to display that intriguing icon prominently, providing evidence of your site's popularity. The current state of play, though, is that unless you've got a really unique idea, resource, database, or what have you that's of incredible cosmopolitan interest, you're unlikely to feature in these magazines, or lists. We'd venture that if you're reading this book, then you're not (yet) in that class of webmaster.

Are these downplaying comments a case of our sour grapes, or a calm statement of fact? Well, it's true, *our* sites have not been featured in any of these electronic magazine type of announcements, so far as we know, but then as working scientists we're rather more used to a sort of low-level, getting on with the job kind of steady as she goes exposure to the public eye. So in fact we guess that we're more appreciative of a constant steady trickle of interested visitors (the gourmets?) than of a waterfall of ravenous hordes (the gourmands?).

Newspapers and Hard Copy Computer Mags:

One of us (test: we told you who earlier on) has in fact had his home page URL listed in the computer section of a national newspaper. Erstwhile fame, eliciting all of perhaps a dozen accesses. Not too impressive, but then it was just a wee letter to the editor, after all. But never mind, you yourself might get your site mentioned in the science section, someday, just like our David might get his long-term analysis of nematode growth in the pages of *Nature*. Our point is, of course, that if David does get his work into say *Science*, and then receives further attention from the science correspondents of a national newspaper, then he'd be a fool not to have a web site ready for perusal by those interested in further components of the story, to which the newspaper piece can direct them.

DO YOU WANT TO RETRIEVE INFORMATION FROM PEOPLE WHO VISIT YOUR SITE?

You might be in for a shock, if you're expecting lots of feedback from www browsers. People like us around the world do tend to visit quietly, but only very rarely, in our experience, do they bother to say howdy-do. Nevertheless, it's always a good idea to give those browsers a chance to get into easy contact with you, and the easiest way for you to expedite this contact is by setting up a hotlinked anchor as a 'mailto' action. Let's see, why don't we make this really easy.

(a) The 'mailto' Option:

The 'mailto' anchor is simple. Instead of specifying the 'http://servername. location.country' URL, when you're placing anchors in your web pages, you specify 'mailto:emailaddress@place.type.country', as in, for example, 'mailto:odonnell@sasa.gov.uk' or 'mailto:larry.winger@ncl.ac.uk'. That's it, really. When a foreign browser clicks the link, through their browser, the 'mailto' option specified in the link will activate the mailer programme in their browser, and they'll be presented with a mailing box and the address (yours) already inserted. The example shown below also illustrates how you can attach a file to your mail through Netscape.

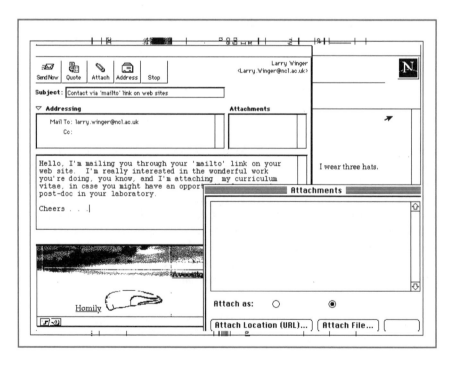

So why don't you can check things out for yourself, just to see if you can email yourself through your browser. Of course, we're going to assume that you've set up your settings on your browser correctly, so that (a) the mailer in the browser software knows your own return address; and (b) the mailer knows your mail server address. Otherwise, your mailings through your browser are not going to work.

This 'mailto' link is quite a worthwhile exercise in itself, in any mailing programme you have, just to see if your settings are all correct. I don't know how many times I've had mailings from people who haven't set their return address correctly. Sometimes I get several plaintive queries, to the effect: 'why don't you reply?'. And of course, reply is impossible, since their return address is incorrectly specified. These poor souls should try, just once, sending a message, using the reply option, to themselves.

(b) Forms on the WWW:

When you get really brave, and want to probe the deep recesses of your browser's psyche, as well as to develop a really useful database on some important matter (maybe you're interested in hirsutism in net surfers, we don't know), you can always try to 'set up a form' on the WWW. Yes, it can be done, and it's even possible to get the returned forms into a database. No doubt you'll want to enlist the services of a real computer person, but if you're a glutton for DIY, and you've got a spare week, and the sort of project that warrants this kind of analysis, then why not give it a go? Seriously, forms are not at all difficult to set up with the new html editors, but there are some snags along the way, and you'll probably hear about some problems from frustrated browsers whose particular software doesn't quite work on your lovely form. Then you'll have to try to adjust. We don't want to say anything about the scientific validity of any study conducted in this way on the net, we just want to note that forms can be set up, and data retrieved from them into databases.

Here's an example of the process I (LW) used to elicit information about the opinions of people on a proposed public artwork out in the North Pennines.

First I composed an opinion form, that I hoped would identify various interest groups among my browsers, which I could attempt to correlate with opinions. In fact, I composed the form directly with the html editor, which allowed me quickly and conveniently to specify popup menu options for the various choices my respondents might make. No doubt you've seen forms on the www by now, but for further illustration, here's a part of the form I composed:

Location: http://georgia.ncl.ac.uk/Opinionform2.html

| What's New? | What's Cool? | Handbook | Net Search | Net Di |

Art appreciator
Cyclist
Walker or rambler
Nature lover
Student
Responsible citizen
✓ **Philistine**

1. Do you know what the issues are?   I am not informe

2. Interest Group: Which label most nearly describes you?

3. Age 70-100   4. Gender Android

5. C2C experience: I will probably not visit the C2C ever.

6. How would you most likely travel to the Skyspace?   I would not bother going to see it.

7. Would the Skyspace be a reason for visiting the area?   I do not know.

8. I think that the Skyspace structure would: have little effect on the environment.

9. I think that the Skyspace structure would: have no effect on tourism or community

10. I think the Skyspace structure would be:
    neither fish nor fowl

Now as I composed the form on the html editor, I was also able through the editor to specify an action for the the foreign browser to take when they click on the 'submit' button:

Document Title: Opinion Form: The Skyspace proposal for the North Pennines

Form Action: mailto:larry.winger@ncl.ac.uk   POST ▼

and I receive the form back in my mailbox, as an attachment. So far, so good. Except the form looks like this:

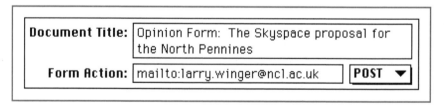

```
=Usenet+Newsgroup&=uk.education.teachers&=Larry+Winger&=Hexham&=Northumbe
rland&=larry.winger@ncl.ac.uk&=I+have+read+the+clippings+and+letters.&=Responsib
le+citizen&=4050&=Male&=I+have+visited+the+cycle+route.&=By+foot&=Yes&=enhan
ce+the+environment.&=enhance+tourism+and+enhance+the+community&=complementar
y+to+the+postindustrial+and+agricultural+heritage+of+the+area&=have+an+artistic+expe
rience.&=I+would+vote+for+the+Skyspace+artwork&=I+think+the+Skyspace+proposal
+could+be+really+a+great+artistic+experience.&name=Submit
```

So I open the form up in a word-processing application, and apply a few global changes, and in a moment or so it looks like this:

> The grapevine Larry Winger told me Larry Winger Hexham Northumberland
> larry.winger@ncl.ac.uk I have read the clippings and letters. Philistine
> 40-50 Male I will probably not visit the C2C ever. By bicycle Yes
> enhance the environment. enhance tourism and enhance the community neither
> fish nor fowl have an artistic experience. I would vote for the Skyspace artwork
> This is the space for comment chump|

Well, that's pretty neat (no doubt astute readers will have noted that I used a different starting form, in the examples above, but that was just to keep you on your toes; the principle stands!) and since I was careful to specify the tab delimited spaces that my database software needs to separate items, now I can import this ASCII text directly into my database programme:

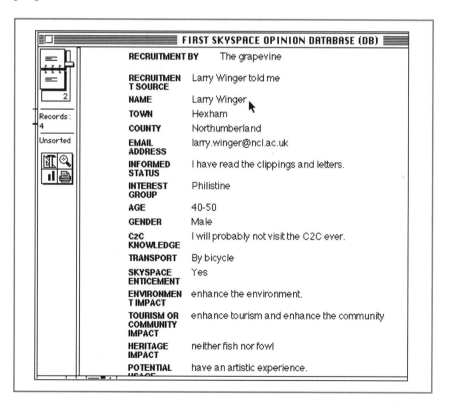

FIRST SKYSPACE OPINION DATABASE (DB)

RECRUITMENT BY	The grapevine
RECRUITMENT SOURCE	Larry Winger told me
NAME	Larry Winger
TOWN	Hexham
COUNTY	Northumberland
EMAIL ADDRESS	larry.winger@ncl.ac.uk
INFORMED STATUS	I have read the clippings and letters.
INTEREST GROUP	Philistine
AGE	40-50
GENDER	Male
C2C KNOWLEDGE	I will probably not visit the C2C ever.
TRANSPORT	By bicycle
SKYSPACE ENTICEMENT	Yes
ENVIRONMENT IMPACT	enhance the environment.
TOURISM OR COMMUNITY IMPACT	enhance tourism and enhance the community
HERITAGE IMPACT	neither fish nor fowl
POTENTIAL USAGE	have an artistic experience.

Records: 4 — Unsorted

A simple matter, in actual practice, and with a little time perhaps I will build up a picture of the opinions of people who are interested in the issue of public money and public art.

You could easily do the same from your own computer, but in practice this sort of approach is labour-intensive and slow. Far better either to take on an advanced web designer as part of your project, or to familiarise yourself in-depth with some of the dramatically functional software packages which will rapidly construct your desired form and convert received information into database-ready data, all without you having to act intelligently in any way. We'll be reviewing the approaches of some of these advanced software packages in our web-site which we are designing to enhance this book, so we'll welcome your visit to these growing pages at:

http://www.compulink.co.uk/~embra/ifs.html

FURTHER INFORMATION ON WEB PAGE COMPOSITION

We mentioned at the beginning of this chapter that we'd let you know where in the web to go to get special tips on web page composition, or 'construction', as we tend to think of *building* a home page. So here's a wee table of useful sites to visit to get some handy tips, useful resources, and just about anything you need, really, to build a really useful site.

TABLE OF USEFUL WEB COMPOSITION SITES:

Self Analysis: Is your Web composition everything it should be?
Look at it yourself, using your own browser. Does it measure up to your own expectations?

Helpful hints
http://www.glover.com/

More advice and resources from Lon Koenig
Lon Koenig at http://www.schnoggo.com/

A comprehensive analysis of web design and construction can be found at
http://web.canlink.com/webdesign/nl.htm

More information on web design can be had through our own
University of Newcastle's computing service
http://www.ncl.ac.uk/~namm2/talks/

The Tea Break

Well, if you've read this book you'll now be busy putting the Internet to work for you. However, if you really want to have colleagues talking in hushed tones about your new-found computer prowess, then we're afraid that you have to become more of a recluse. Stop consorting with others at tea break time (it'll give them more opportunity to talk about your new-found computer prowess and how anti-social you've now become) and start having your tea break with your computer instead.

So, position that Gary Larson mug where it can't fall onto the keyboard and let's see what fun things you can find on the Internet when you're not working.

Well one thing to do is visit the on-line versions of those old tea break standards *New Scientist* (http://www.newscientist.com) and *Nature* (http://www.nature.com).

Being a conscientious scientist you might also want to brush up on the facts you need to refute those silly animal rights arguments that someone in the pub comes up with — so you want to head for the facts at the Research Defence Society's WWW pages at http://www.uel.ac.uk/research/rds/

However, perhaps you'd like a break from science altogether. Something more arty perhaps? So a tour of the world's great works of art at the Louvre (http://www.louvre.fr) or indeed the Rene Magritte pages at http://www.virtuo.be. Hint: impress your friends by using one of the illustrations from these sites as your windows wallpaper.

Well high culture is fine but the occasional splash of humour doesn't go amiss. A visit to the Monty Python (http://www.gbar.dtu.dk/~c938202/pythonpage/python.html) or Father Ted (http://www.geocities.com/Paris/2694/craggy.html) pages will soon cheer you up, or perhaps a visit to the Doonesbury site http://www.doonesbury.com.

If you fancy something more interactive than web surfing then the whole of Usenet is your oyster with Dejanews (http://www.dejanews.com). If Professor Realitas has limited the newsgroups received on your PC to the science ones, you can still read and contribute to them all at the Dejanews site. Remember to add that all important disclaimer "I am not speaking on behalf of my department/company/Professor Realitas" before adding your outspoken contributions.

Tea breaks are also a good time to appear knowledgeable about security. One of your colleagues will undoubtedly triumphantly inform you (usually with a grotty photo-copy) of the horrible viral perils that await those poor souls who happen to open an email entitled 'Good Times'. You'll be able to smile nonchalantly and inform your crestfallen colleague that this computer virus hoax is one of the oldest, and most wide-spread, on the net. Have a look, you'll say smugly, at the most enlightening Internet Virus-Myths site (http://www.kumite.com/myths) where virus hoaxes are identified and counter-measures are suggested. There's also a lot of links there to really useful sites that explain viruses and virus hoaxes, like Computer Knowledge (http://www.cknow.com), Stiller Research (http://www.stiller.com) and Seven Locks Software (http://www.sevenlocks.com). Rob, webmaster at the kumite site, has a wonderful way of exhibiting any possible bias, and internally validating his approach.

Security, of course, is a problem that must be recognised and dealt with, except for those who are not connected, and never use any floppy discs that have been out of their sight. It's a truism, isn't it, that hermits rarely catch colds! Of course, the safest personal computer is one that never gets exposed to any possible danger. And this approach is practiced, quite literally, by institutions like pharmaceutical companies, who supply their sales representatives with laptops, and who permit useful interaction in their Local Area Network, but not out in the big world. At all. Full stop. Period. Fine, understandable.

But where's the fun in that? You wouldn't be reading this book if you didn't want to be involved. Using your own lap-top, we presume, as opposed to your company's, if you happen to be under those restrictions. While the dangers of downloading an unwanted virus (are there any that are wanted?) from rogue FTP sites are commonly recognised, and are dealt with in the same way that you'd deal with a foreign disc – ie. by running the file through a virus-checker and sanitiser program! It's also important to recognise that you expose yourself and your computer to a small possibility of danger just by being connected to other computers. One example that's currently making the rounds is the unforeseen role of Netscape's 'cookies', pieces of information about your WWW browsing, complete with your identity, that the browsed site can echo to identify you, and even to include you or your computer on highly idiosyncratic mailing lists. Another good way to become exposed, sadly, is to post a comment on the newsgroups. Just like in the real world, it seems, security is always something of a risk, when you venture out of doors.

But this security thing can become too much of a needless worry, we think. It's a fact of life that most academic institutions around the world list their staff email addresses. Not that they're published as a big list, of course, but rather they're available if you know the particular person's name, usually by a gopher search. If such a thing as the 'Good Times' email virus were really feasible, you'd hardly expect university computing services to lay themselves open to such attack, now would you? Oh go on, we agree, just because one is paranoid doesn't mean they really aren't out to get you.

So what *not* to do on your tea break is to leave your PC logged in and unattended, or leave your password written on a post-it beside the screen (just in case you forget it). If you do, you could well find yourself in the unfortunate position of the man who found that thousands of messages offering child pornography for sale had been sent all around the world using his account.

If you are involved in espionage, intra-departmental back-stabbing, an affair, or are just plain paranoid, you will be alarmed to find that, theoretically, your email can be intercepted at any of the various computers through which it passes to reach its final destination. For this reason there is an increasing interest in encryption software which encodes your messages so that only the intended recipient can read them. The best known example is that of Pretty Good Privacy, or PGP as it is universally refered to. This software's encryption is allegedly uncrackable. Find out all about it and get a copy (it's Freeware) at http://www.ac.uk.pgp.net/pgp/ukerna.html.

Another useful way to spend your lonely tea-break is to attend to the mailing lists and web pages that you might be responsible for. And if your tea-break time runs on into work time, well, after all you're contributing to the growth and dissemination of scientific information, which is your job, really, isn't it! Though you may have trouble convincing your academic paymasters that your work is project-related. But what the hey, it's tea-time. It can be especially useful, however, as you sit at your computer gazing into cyber-space, to know how to 'eliminate' any window at a quick key-stroke!

Tea-time is a useful convenience, after all, to run special searches for a friend, colleague or neighbour. My folk music friends always want to know the words to this or that song, so I either check out the Digital Tradition folk-song database at its new home at http://www.deltablues.com, or post a query on the uk.music.folk or rec.music.folk newsgroups.

Do your colleagues discuss their holiday plans during tea-break? You'll want to impress them with your vast holiday expertise, and your ability to book your vacation through the Internet. Don't forget to consider the Elpha Green Cottages holiday-let in the balmy North Pennines! Too bad you can't take the trains from Ulm <<http://rr-vs.informatik.uni-ulm.de/rr/>>. But seriously, any holiday venue you might want to consider can be found so easily using the big search engines.

Tea-breaks are a good time to brush up on your Internet skills. We'd be remiss if we didn't mention (maybe we already have?) the Netskills pages at: http://www.netskills.ac.uk a national project site hosted by the Computer Services here at Newcastle University.

On the rare occasions when you join your colleagues for tea, you will be required to impress them with tales of the latest happenings on the web. Luckily, there is an organisation which will provide you with up to date information absolutely free. The Netsurfer's Digest is a regular bulletin of the latest interesting WWW sites and other net-related info. You can subscribe by sending an email to the address: nsdigest-request@netsurf.com. The message to send is either subscribe nsdigest-html or subscribe nsdigest-text, depending on whether you would like to receive the digest as a text or html document. Tip: html is better. You will now be regularly kept up to date with what is happening. You will also be a member of that select club that raises a wry smile every time they read about a new WWW site in one of the newspaper columns now devoted to the subject. As often as not, their sites are lifted from the latest Netsurfer's Digest.

* * * * *

Well, thats it from us. We hope that this book helps you get a lot out of the Internet, both professionally and personally. We wish you more productive communications, enlightened access to similar experts around the world, and better and better science! Happy surfing!

Glossary

"What does that word mean again? I'm sure they used it a few pages back. Oh hold on . . . it'll come back to me in a minute . . ."

Look, don't waste time, just use this glossary — that's what the publishers made us write it for.

Access:

Verb or noun indicating that you've reached into another computer and retrieved information.

Archie:

A system for finding files available on FTP servers. This is the sort of thing that people used before they invented the WWW and AltaVista. Sometimes found in conjunction with Anarchie, which is a piece of software that uses Archie.

Attachment:

A file which is attached (hence the name!) to an email message.

Bandwidth:

A term which is often used in relation to the amount of data the Internet can take — e.g. "Video files on WWW sites use up too much bandwidth." If pressed for a definition however, people become very vague and look at their feet. What it actually means is, well, it's the amount of data that an Internet link can handle at one time. Or something.

Browser:

The software you use to access the WWW. Synonym of 'Netscape'.

BTW:

By the way. One of the many acronyms found in Email and Usenet messages. Tip: If you find yourself slipping these in to real speech, it's time to get out more.

Client:

Between computers on any network, there seems to be this thing called a client-server relationship. One computer is the client, and the other is the server. We think it's a function of who's on top.

Domain:	A part of an email address or of a computer's Internet address.
Downloading:	The act of transferring a file from somewhere on the Internet into your own PC. Hint: "I'm just downloading some data" sounds so much more impressive than "I'm just aimlessly wandering about the WWW" when your boss sticks his or her head round the door.
Email:	Electronic mail. Look if you're having to look that up in the glossary, just go back to the beginning and start again OK?
Ethernet:	The bits of cable that connect PCs together in a local network or, indeed, the local network itself.
FAQ:	Frequently Asked Question; rhymes with yak, but don't let the Q say its name.
File:	Could be anything really — a picture, a document, a web page, a programme. They're all files.
Flames:	Abusive Internet messages. Of course this derogatory term only applies to other people's insults and not your own well honed Edmund Blackadder-isms.
Freeware:	Software that is available for free. Yes really — and some of it is very good too.
FTP:	A way of retrieving files from the Internet and also for loading files onto your WWW page. It stands for File Transfer Protocol. And nothing else — even if you do live in Central Scotland. Confusingly, it can refer to both the method of retrieving files and the software that does it.
Gif:	A popular format for image files. Hence people tend to talk about image files as 'gifs' (or 'jpegs', the other common format).

Gopher:	Gopher was what people used to find and retrieve information from other computers before the WWW was invented and made life so much simpler. Still seen sometimes when you access a site and the gopher there asks for some information, or a search term. Otherwise an endangered species.
Home page:	Your own page (or the front door of anyone, institution, or facility) on the WWW.
HTML:	The language that WWW pages are written in — basically English with a few tags put in here and there. Hyper-Text Mark-up Language, since you ask.
Hypertext:	Documents that contain links to other documents.
IP:	Internet Protocol — how information is sent across the Internet to reach a specific destination.
IRC:	Internet Relay Chat. CB radio for the 90s. Except that people use names like 'cyberphreak' instead of 'rubber duck'. You get the gist though: it allows people to send messages to each other in real time.
ISP:	Internet Service Provider. The company or institution that provides you with an email address, WWW space and a link to the Internet.
Jpeg:	See Gif.
Links:	This catch-all term just refers to the connection from one WWW page to another.
Listserv:	The most common type of software used to run mailing lists. It's almost become a generic term for mailing lists.

Mailing list:	An email-only discussion group, where messages sent to a list address are forwarded to everyone subscribed to the list.
Moderator:	Someone whose job it is to approve the postings to a 'moderated' mailing list or newsgroup. A moderated mailing list or newsgroup is one in which any messages are first approved by a moderator. We hope that's clear.
MUD/MOO:	Sort of IRC with attitude.
Newsgroup:	An electronic conference on Usenet devoted to a particular subject, and to which you can send, read and reply to messages.
Newsreader:	The software you use to read and send Usenet messages.
Post:	A message sent to a Usenet newsgroup (either noun or verb form), the person who sends it being the poster. Now, what we want to know is if the thing that you put up at conferences is called a poster, what do you call the person who puts the poster up?
PPP:	Point-to-Point Protocol. This allows your PC to become part of the Internet and send and receive information.
Remote computer:	We use this a lot, so we thought we'd better put it in. It just means a computer that isn't the one in front of you.
Server:	A computer that receives and distributes Internet traffic.
Shareware:	Software available on a 'Try before you buy' basis. The Internet equivalent of an 'honesty box'.
Sig:	A signature that you put at the end of your email messages and Usenet posts. Typically it

will have your name, a witty quote ("Molecular biologists do it in base pairs" or something equally mirth-inducing) and a statement making it clear that you are not speaking for your employer nor for your maiden aunt.

Site: A collection of related pages on the WWW.

Spam: Widely and inappropriately distributed email and Usenet messages. It will not be too long before the words "Make Money Real Fast" appear on your newsreader or in your in-tray.

Sysadmin: The person who runs a local network.

TCP: Transmission Control Protocol. Often found in the company of IP, this is one of the things that allows computers to send and receive information.

Telnet: A way of connecting to and using a remote computer. As with gopher, the WWW makes life so much easier.

Thread: A string of Usenet messages (posts) or even sequential mailings on the same topic (the RE: subject heading keeps the thread accessible for archiving purposes).

Uploading: Transferring files or messages from your computer to another one. Hint "I'm just in the middle of uploading some data" sounds much better than "I'm just in the middle of firing off a contribution to that fascinating thread on favourite vegetarian recipes", yet they both mean the same thing.

URL: A URL is the address of a WWW site or document. What does it stand for? Do you really want to know? Oh alright then, it's Uniform

	Resource Locator — but we had to go and look it up in another book's glossary. That's how important the full name is.
Usenet:	A collection of electronic bulletin boards where people swap insults, rumours, pictures of naked men and women and, occasionally, useful information.
WWW:	World Wide Web. Though if you think about it, that doesn't make sense. Surely it should be World-wide Web? But then shouldn't it be abbreviated to WW? Best not to ask.

If the particular piece of jargon you were looking for an explanation for isn't here, then you can always try emailing us: larry.winger@newcastle. ac.uk or odonnell@sasa.gov.uk.

Index

DATE DUE

The Library Store #47-0204